OCEAN WAVES
AND
KINDRED GEOPHYSICAL PHENOMENA

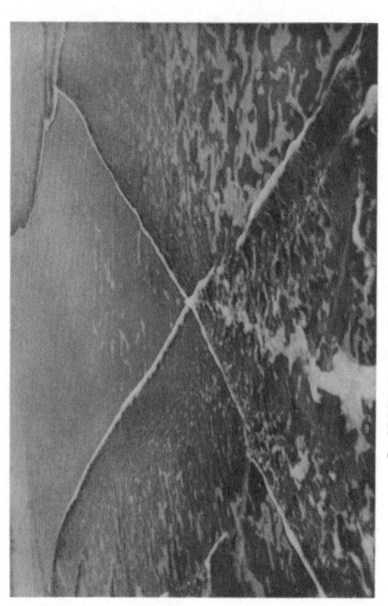

I. CROSSING WAVES IN SHALLOW WATER
(Branksome-Chine near Bournemouth)

OCEAN WAVES

AND

KINDRED GEOPHYSICAL PHENOMENA

by

VAUGHAN CORNISH, D.Sc.

With photographs by the Author

AND

ADDITIONAL NOTES

by

HAROLD JEFFREYS, M.A., D.Sc., F.R.S.

CAMBRIDGE

AT THE UNIVERSITY PRESS

1934

CAMBRIDGE
UNIVERSITY PRESS

University Printing House, Cambridge CB2 8BS, United Kingdom

Cambridge University Press is part of the University of Cambridge.

It furthers the University's mission by disseminating knowledge in the pursuit of
education, learning and research at the highest international levels of excellence.

www.cambridge.org
Information on this title: www.cambridge.org/9781107559998

First published 1934
First paperback edition 2015

A catalogue record for this publication is available from the British Library

ISBN 978-1-107-55999-8 Paperback

To

HERBERT JOHN FLEURE
D. Sc.

IN TOKEN OF
REGARD AND ESTEEM
I DEDICATE THIS
BOOK

CONTENTS

PLATES

x

ILLUSTRATIONS IN TEXT

PREFACE

This narrative of nature study contains an account of the author's original contributions to the knowledge of the waves which are raised by wind upon the ocean, of the kindred forms which wind and currents raise and propel in sand and snow, and of tidal bores and other progressive waves in rivers which travel under the action of gravity alone.

The research originated under the following circumstances. In the early 'nineties, after searching the South Coast for a place of residence, I found a beautiful, and at that time secluded, spot on the cliffs near Branksome Chine, between Bournemouth and Poole Harbour, and living here within a stone's throw of the shore, I was mastered by the fascination of the waves that broke in ever-changing cadence on the beach. Near by, moreover, a little stream making its way through the beach threw the sandy bottom into curious undulations which actually travelled up-stream against the current by which they were formed; and a mile or two away on the dunes near Poole Harbour the wind drove the dry sand in wave processions.

My time was my own, and I decided to investigate these various wave phenomena. In the course of a few years, however, I was confronted with a disagreeable alternative, for I had either to leave the research incomplete or give up my beautiful home and travel widely in search of waves. I decided on the latter course, and the present volume gives a summary of the results of my observations both in England and abroad. I have never again had a house with such a view as that from my former home on the cliff, but I have compensations in the memory of many wonderful sights in storms at sea; of snow-waves moving in ghostly procession across the Canadian prairie; of sand-waves, rank behind rank, driven by the desert wind; of the onset of the tidal bore in the Severn and the Trent, and of Leaping Waves in the Rapids of Niagara.

In one respect only was my work, for a time, disappointing. Part of my purpose was to provide the mathematician with numerical data for the further development of the theory of water-waves, and to enlist his interest in the progressive undulations of granular material. But there are fashions in mathematics as in all human pursuits, and I had to wait no less than thirty years before a younger generation of mathematicians, more interested in "turbulence" than their classical Victorian predecessors, began to develop the material which I had collected. Foremost among these was my friend Dr Harold Jeffreys, whose "Additional Notes" at the end of this volume generalise and extend the results of my observations and measurements.

VAUGHAN CORNISH

Inglewood, Camberley, Surrey
14 *November* 1933

BIBLIOGRAPHY

Selected list of original contributions by the Author upon Waves and kindred phenomena.

1. "On the Formation of Sand-Dunes", *Geogr. Journ.* March 1897.
2. "On Desert Sand-Dunes Bordering the Nile Delta", *Geogr. Journ.* Jan. 1900.
3. "On Sand Waves in Tidal Currents", *Geogr. Journ.* Aug. 1901.
4. "On Snow-Waves and Snow Drifts in Canada", *Geogr. Journ.* Aug. 1902.
5. "Progressive Waves in Rivers", *Geogr. Journ.* Jan. 1907.
6. "The Jamaica Earthquake (1907)", *Geogr. Journ.* March 1908.
7. "On the Cause of the Jamaica Earthquake", *Geogr. Journ.* Sept. 1912.
8. "Ocean Waves, Sea Beaches and Sandbanks", *Journ. Roy. Soc. Arts,* Nov. 1st and 8th, 1912.
9. "Observations of Wind, Wave and Swell on the North Atlantic Ocean", *Q.J. Roy. Met. Soc.* April 1926.
10. "Waves in Granular Material Formed and Propelled by Winds and Currents", *Monthly Notices* of R.A.S., Geophysical Supplement, July 1927.
11. *Waves of the Sea and other Water Waves.* Pub. T. Fisher Unwin (London, 1910), 8vo, 374 pp., with 50 photographs taken by the Author.
12. *Waves of Sand and Snow and the eddies which make them.* Pub. T. Fisher Unwin (London, 1914), 8vo, 383 pp., with 88 photographs taken by the Author, 30 diagrams and 2 maps.

CHAPTER I

THE SIZE AND SPEED OF OCEAN WAVES IN RELATION TO THE VELOCITY OF WIND

(1) *Measurements from a ship hove-to during a storm*

The North Atlantic was very stormy in December 1911, and when the P. and O. liner *Egypt* passed Gibraltar homeward bound, the ocean was heaving with a long, swift swell. On the morning of the 19th, between Cape St Vincent and Cape Roca, the ridges occasionally rose above my line of sight when I was sitting out on the promenade deck, the eye-height being then 23 feet above the water-line of the ship.

Watching a small tramp steamer as it was raised and lowered bodily by the swell, I was struck by the fact that when she was on the top of the billow the water lay almost level along her side. This at once suggested to me an explanation of the discrepancy which I had often noted between the wave-lengths calculated from the period (or interval of time between the passage of the crests) and the estimates from the apparent position of the crests along the side of the ship on which I was travelling. My own estimates, and also those of navigating officers with whom I compared notes, were quite out of accord with the calculated length, being invariably much less than the latter. It now occurred to me that the eye was deceived, that attention was fixed upon the steep shoulders of the advancing and receding waves, and that the considerable part of the wave-length comprised in the greatly foreshortened, nearly flat, top had escaped observation.

I did not have to wait long for a confirmation of this idea, and also for the discovery of an additional source of error. Next day, December 20th, we passed Cape Finisterre, and, changing course somewhat, headed straight for Ushant across the Bay of Biscay. The heavy swell was still from the same direction, and since it was more exactly abeam, I was able to determine the true period without need of correction for the speed of the ship. The interval between the arrival of the rollers was 11·4 seconds. According to the mathematical theory of deep-sea waves the speed in miles per hour ("statute", or land-miles, not geographical or nautical miles) is obtained by multiplying the period in seconds by $3\frac{1}{2}$, so that a swell with a period of 11·4 seconds has a speed of 39·9 miles per hour. The wave-length or distance from crest to crest, reckoned in feet, is according to theory, equal to the square of the period multiplied by $5\frac{1}{8}$, so that the wave-length of this swell would be 666 feet.

The height above the ship's water-line was commonly 20 feet, and occasionally a few feet more, and as the ship lay exactly along the trough of the waves and therefore sank to her proper floating level, this was the full height from hollow to crest.

The breeze blew from the south-west, a direction

2

nearly at right angles to the line of advance of the swell. Early in the day, being light, it did not interfere with the pattern of the ridges, but freshening later, formed fairly steep waves, which, cutting up the ridges into short lengths, so camouflaged the swell that the sea no longer presented the appearance of undulation upon a grand scale.

During the night the wind changed direction and blew from west-north-west, rapidly increasing in strength until in the small hours of the morning of December 21st its force was beyond the common experience of North Atlantic gales. Several ships sank with all hands in the Bay, and our liner was for a time in peril, until brought round with great difficulty so as to face the waves which had struck her abeam. With just enough steam to hold the bows head-on to the sea, the ship was kept hove-to until one o'clock in the afternoon of this memorable day. The opportunity for which I had longed for years had arrived, and at last I had a stationary post of observation amidst the great waves of the ocean. Moreover, by a fortunate and rare coincidence, the tremendous wind had come on to blow exactly in the direction of the heavy swell already running, and so there was no confused turmoil, but one magnificent procession of storm-waves sweeping across the sea from horizon to horizon. The sky cleared, and between the chasing clouds the sun shone down in strength, for we were still no further north than the latitude of La Rochelle.

At 8 a.m. as I stood upon the promenade deck, with an eye-height 27 feet above the water-line, each passing wave was well above the horizon. From the look of the sea I judged the waves to be as high as any which I had seen on previous voyages, that is to say, 40 feet from trough to crest. They were remarkably uniform in height and were much steeper in front than at the back, thus differing from the almost symmetrical swell of the preceding day. The velocity of the wind at this time, according to the seamen's estimate, was 52 miles per hour, somewhat greater than the "Strong Gale", expressed by the number 9 on Admiral Beaufort's scale of force.

From my station on the promenade deck, I timed the arrival of successive wave-crests and found that the "period", or interval between them, was 13·5 seconds. The theoretical wave-length for this period is 934 feet. By 10 a.m. the waves looked lower and their period had decreased to 12·5 seconds. For this period the calculated wave-length is 801 feet. Nevertheless, as I looked fore and aft at the water's brow advancing and receding, the wave-length seemed at times rather less than the length of the ship, at other times rather more, and the length of the ship was only 512 feet. I judged the height of this watery sky-line near the bow to be about 4 feet above my line of sight, that is to say, 31 feet above the floatation line. The discrepancy as to wave-length is explained by the fact that the slope of the wave is steep in the first half of the upper portion but becomes very slight nearer to the crest, so that if the elevation of the observer's point of sight amidships remain constant whilst waves which exceed this altitude travel past the vessel from stem to stern, more and more of the crown of the wave will be

4

eclipsed without any warning of the occurrence to the eye. Meanwhile the flickering brow of water becomes more spectacular and impressive. I am of opinion that in these circumstances it is usual to take as the true crest of the wave a point very far short of the actual summit.[1] The flatter the wave the greater the error in judging wave-length from such a standpoint, but the under-estimate of height is relatively small, and thus there results a tendency to over-estimate the *steepness* of waves during storms of exceptional violence when, as will be described later on, the waves are flatter than during moderate gales, although of greater height. The tendency to a systematic over-estimate of steepness is serious from the standpoint of the naval architect, who has to calculate the greatest strains resulting from variation of buoyancy in wave-water to which a ship is liable in exceptionally stormy weather.

Desiring to view the scene from a loftier and more open position I applied for permission to observe from the navigating bridge, which the Commander, Captain F. R. Summers, kindly granted, and I went up to this station soon after 10 a.m. and remained there until 1 p.m., when the ship was again put on her course.

Upon the bridge, the level of the eye was 54 feet above the ship's water-line. This was well above the crests of the waves, which were, moreover, decreasing. The position commanded a bird's-eye view of the whole length

[1] An excellent illustration of this effect is provided when motoring on a steeply undulating road. From the top of one ridge we see the true summit of the next but, as we descend, the *brow* of the next advances towards us.

of the ship. By noon, when the wind had dropped to 40·25 miles per hour (according to the estimated force of the Beaufort scale), it was evident to the eye that although the waves had been decreasing in size throughout the morning, they had even now a wave-length considerably greater than the length of the ship, for when the stern of the vessel was lifted on one crest the bow was dipped, and the next crest could be seen approaching the ship. At this moment I estimated the distance between the bow and the approaching crest to be 100 feet. The Commander, who joined me in these observations, judged the interval to be 30 feet, so that his estimate of the wave-length was 542, mine 612 feet. The period of the waves at this time was 11·27 seconds, which according to theory is that of waves 651 feet in length. Thus from a standpoint high above the crests, the wave-length judged by eye is in fair agreement with that calculated from the period. This result shows clearly that the enormous discrepancy between the wave-length calculated from the period and that estimated when the observer is below the level of the crests is due to those deceptions of the eye which I have already described. Thus for a reliable eye-estimate of wave-length the observer should take up the highest position which the structure of the ship allows, whereas for a reliable estimate of height he should not be above the tops of the waves. Unfortunately it is very seldom that the estimate of the size of waves, i.e. of both height and length, is recorded by any experienced observer except an officer who is tied to one position by duty on the bridge.

6

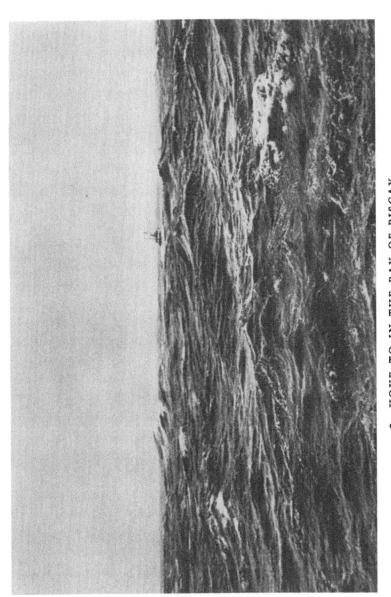

2. HOVE-TO IN THE BAY OF BISCAY

The theoretical speed of deep-sea waves of the period recorded at noon, 11·27 seconds, is three-and-a-half times this figure, that is to say, 39·4 statute miles per hour. From the view-point of the bridge, the ship being held steady against the sea, I obtained a direct measurement of the speed of the waves by noting the time taken by the crests in running from stem to stern, and the speed so determined was 43 miles per hour. This concordance of observed and theoretical speed is valuable, as showing that a formula hedged round by limiting conditions which the mathematician is careful to specify but the layman apt to forget is applicable to the conditions of a storm at sea.

The outstanding impression left upon my mind by this rare opportunity of observing storm-waves from a ship hove-to was the small difference between the speed of wind and speed of wave, for throughout the morning there was never more than a "Slight Breeze" (number 2 of Beaufort's scale), blowing over the ridges of the great waves.

I was also impressed by the fact that no independent heave, or swell, of the sea was visible. The fact that the wind blew in exactly the same direction as the swell already running would account for the absence of a crossing swell. The suddenness with which the sea rose was evidently due to the fact that a wind of greater speed and identical in direction developed the swell already running, increasing both its height and speed. But in text-books and treatises it is customary to compare the swell at sea to the longer waves, swifter and more enduring, which emerge from the disturbance caused by a stone thrown into a pond, with accompanying references to "Fourier's theorem". This treatment appears to imply that one should expect the presence of a longer heave of the sea in the direction in which the wind is blowing, moving with a speed greater than that of the wind. When I began observing waves, in 1895, this expectation was fostered by two circumstances, now no longer obtaining. The first of these was the prevalent under-estimate of the length of the dominant waves at sea, due to the effects of foreshortening and eclipse which I have explained above. The second was the erroneous factor for calculating the speed of the wind from the revolutions of the cup-and-ball anemometer, which gave a speed considerably less than that now calculated. Thus on account of two sources of error, it was supposed that the waves driven by wind moved more slowly than is actually the case and that the swell frequently outpaced the velocity of the wind to which it owed its origin. In the succeeding sections I shall describe observations which indicate that the wind only exceeds the final velocity of the wave by a small fraction, and that no swell visible at sea, or capable of forming breakers on the shore, exceeds the speed of the wind by which it was produced.

(2) *Observations at sea on the effects of squalls upon waves*

On December 22nd, 1906, in Mid-Atlantic Lat. 30° 21' N., Long. 35° 43' W., on a south-westward course from Liverpool to Mona Passage, a following wind, in force

8

between a Strong Breeze and a Moderate Gale, had raised considerable waves. At 4 p.m. a violent, following squall accompanied by rain overtook our ship, which was making 11 knots, and passed over us in 4 minutes. While we were in the squall, the waves around were much larger than those immediately preceding and following it, about 7 feet higher as well as I could judge.

Next day, December 23rd, under similar conditions, a following squall overtaking the ship at 3 p.m. passed over us in 3 minutes. In this case I was able to watch the action of the belt of stronger wind to increase the waves, which, as I judged, grew in height at the rate of at least 2 feet per minute, or 6 feet altogether, the squall increasing the waves which were already running by something like one-third of their height. The frontage, or crest-length of the ridges, was also notably increased. Two minutes after the squall had passed ahead we were in water no rougher than before, but I could see four great ridges travelling on with the squall. At 5 p.m. another squall arrived which passed over us in 5 minutes, and was accompanied by large waves.

Next day, December 24th, our noon position being Lat. 33° 38′ N., Long. 43° 58′ W., the wind dropped at 3 p.m., and by 4 p.m. the sea was slight. At 4.55 p.m. a band of black cloud overtook us, extending across the zenith from horizon to horizon, both ends very dense and apparently discharging rain. A slight additional breath of wind and a few drops of rain were the only disturbances in the air which accompanied the passage of the black cloud overhead, but the sea below was

9

disturbed by a heavy swell, twelve or more great ridges, which made the ship roll heavily. In 5 minutes the cloud had passed on ahead, in another 10 minutes the sea had returned to its former, quiet state.

In order to understand these events we must have some idea of the "fetch" of the squall, that is to say, the breadth in profile of the band of water which it roughened.

In the examples given above it was evident that the rate of advance of the squall was much greater than the speed of our ship, and I will assume for the sake of clearing our ideas, that it was proceeding at the average rate of advance of Atlantic storms, that is to say, at about 30 statute miles per hour.[1] Allowing for the speed of our ship, the relative rate of advance of the squall would be 17 miles per hour, or rather more than 1 mile in 4 minutes, which was the average time taken by the squalls to pass the ship.

The length of the waves on these days being approximately from 300 to 400 feet, the number in a mile would be between seventeen and thirteen, which is concordant with my observation of "twelve or more great ridges".

Extracts from ships' logs which have been kindly supplied me by Dr G. C. Simpson, C.B., F.R.S., Director of the Meteorological Office, show that, in very heavy waters, squalls are sometimes of longer duration. Thus in the log of the S.S. *Graciana* (Captain J. Clark), Glasgow to Boston, U.S.A., December 19th, 1922, Lat. 55° N., Long. 24° 49′ N., during a Strong Gale (Force 9) squalls

[1] See F. J. Brodie, "On the Prevalence of Gales on the Coasts of the British Isles during the years 1871-1900", *Q.J. Roy. Met. Soc.* 1902.

were recorded throughout the day which lasted from 10 to 20 minutes. In the log of the S.S. *Port Stephens* (Captain J. R. Strawbridge), Adelaide to Port Pirie, November 16th, 1923, Lat. 35° 13′ S., Long. 137° 58′ E., in a gale of Force 8 frequent squalls of Force 10 were recorded with a mean duration of about 10 minutes.

Deep-sea waves are group formations, no wave existing by itself. Thus if the breadth of ruffled waters (its profile in the direction of the waves' progression) be not more than the length of a single wave, we approach a limiting condition. I shall go on to show that this limit is not reached in a Squall (which lasts for minutes) but in a Gust, that is a sudden increase of wind which only lasts for seconds.

(3) *Observations of the period of breakers indicating the limiting speeds of waves and the relation to speed of wind*

The winter of 1898–9 was exceptionally stormy in the North Atlantic, and from time to time the ocean swell reached Bournemouth Bay, swinging round Old Harry Rock and breaking in majestic rollers on the sandy beach at Branksome Chine. My home overlooked the Chine, and the path along the low cliff just beyond the garden fence was a perfect observatory for waves. On the morning of December 29th, with an off-shore wind holding back the crest of the wave so that it did not comb over until reaching the shore, groups of great breakers making a slow and solemn booming on the beach arrived at intervals. A group of four breakers came at 11 o'clock, followed soon by another group of four; then at 11.25 a.m. by seven breakers, at 11.35 a.m. by six, and at 11.50 a.m. by the final set of seven. The sea in the intervals between was comparatively smooth. The average period of the waves was 20 seconds. Their speed in statute miles per hour before reaching shallow water ($3\frac{1}{2}$ times this number) was therefore 70 miles per hour.

Early in the afternoon there was an even more re-markable occurrence, the arrival of one hundred and thirty-nine great breakers in succession with no gap or intermission. The regular booming of these breakers continued for 44 minutes, the average period being 19 seconds, corresponding to a speed in deep water of 66·5 miles per hour.

It is suggestive that the swifter waves which came on ahead in groups, occupying individually 1 to 2 minutes in arrival, were spread over a period of 52 minutes, which differs by only 8 minutes from the duration of the main body which arrived later. From the experience of the effect of squalls at sea recorded in the last section I infer that in the Atlantic storm which sent these rollers into Bournemouth Bay the sea had been for the most part in waves which travelled at 66·5 miles per hour, but was barred across with relatively narrow strips of larger, swifter waves where the squalls blew.

The limits of sustained velocity of wind during this stormy period are shown by the following anemometer records: at Alnwick, in Northumberland, on the night of December 27th–28th, 77 miles per hour at 10 p.m. and 71 miles per hour at 2 a.m.; at Fleetwood, in Lancashire, during January, 75 miles per hour for one hour; and

12

at Southport, 73 miles an hour for half-an-hour. In Mid-Atlantic a force of 11–12 on Beaufort's scale, corresponding to not less than 75 miles per hour, was logged on December 29th.

These data indicate that the whole group of waves which occupied 44 minutes in arrival may well have been subjected simultaneously to wind of 75 miles per hour, exceeding them in speed by 8·5 miles per hour, which is a "Gentle Breeze" (number 3 on Beaufort's scale).

We have next to enquire what was the average velocity of the wind to which the 70-mile-an-hour waves were exposed during the squalls. Ordinary records show the average velocity of wind sustained for an hour, and the extreme velocity momentarily reached in gusts, but the average velocity during the squalls is not recorded. I am therefore greatly indebted to Dr G. C. Simpson for the special investigation of this figure from the wind-trace as shown by anemograms for twenty-seven gales recorded in the Scilly Isles. These show that the average velocity during the squalls was 13 miles per hour greater than that immediately preceding the squall.

The opinion of members of the Marine Meteorology Division of the Office was that squalls generally exceeded by two numbers of the Beaufort scale the force of the wind in the lull between squalls. The difference of average velocity of a Moderate Gale, Force 7, and a Strong Gale, Force 9, is 15 miles per hour. It must be noted, however, that the velocity during squalls is included in the average velocity of the gale, so that the excess of speed above the latter figure must be less.

Taking 10 miles per hour as a reasonable value for the excess of velocity during squalls over the average recorded for the gale, the speed during the squalls of the Atlantic storm which sent the rollers into Bournemouth Bay on December 29th, 1898, may be taken as 85 miles per hour. The average speed of the waves in the short groups which came in advance of the rest was, as we have seen, 70 miles per hour, so that on this computation, they would have been subjected during the squalls to the pressure of a Moderate Breeze (number 4 of Beaufort's scale).

The greatest wave-period which I have ever recorded was on February 1st, 1899, at Branksome Chine when a group of twelve large breakers came at intervals of 22·5 seconds, corresponding to a speed of 78·75 miles per hour, the whole group occupying rather more than 4 minutes in arrival, a normal duration for a squall. An average wind-speed of 78 miles per hour has been maintained for a whole hour on one occasion at Fleetwood, on December 22nd, 1894, and in such a gale the squalls would, as I reckon, have an average speed of 88 miles per hour, exceeding that of the swiftest swell of my records by the force of a Gentle Breeze (number 3 of Beaufort's scale).

The momentary gusts registered by anemometers reach velocities far beyond those of waves either as calculated from the period of breakers or from observations at sea. It is important to remember that wind is not merely subject to the slight flicker which is common in quick currents but is essentially a very jerky progression, for the speed of the air rapidly varies nearly, if not quite, in the proportion of 1 to 2.

During the storm of December 6th, 1929, the anemogram taken at Scilly recorded a momentary velocity of 111 miles per hour, the greatest ever recorded in the British Isles. The maximum sustained velocity in this gale was 68 miles per hour. Waves travelling with eight-tenths of the speed of the strongest momentary gust, that is to say, at 89 miles per hour, would have a period of 25·4 seconds, which is much beyond that of any breakers which I have observed.

Although gusts are but momentary, they recur frequently, and I suggest that their ineffectiveness as regards the production of waves, or at all events noticeable waves, is mainly due to their small length of fetch upon the water. A gust travelling with the storm at the normal rate of 30 miles per hour and lasting a few seconds (at a fixed point) only ruffles the water on part of one of the long waves formed in a great storm. It is therefore on too small a scale to re-model the pattern of the undulations of the air and of the eddies in the lee of the crest, even of a group of three or four waves.

As I proceed with my narrative of observation it will, I think, become increasingly apparent that for the proper understanding of the growth and progression of waves at sea we should always endeavour to visualise the eddies and undulation of the superincumbent air. The large scale of the eddy to leeward of the crest, and the very active rotation of the air, are actually visible when wind is propelling waves of granular material such

15

as the desert sand or the dry snow of the Canadian prairie.

I have not had the experience of ocean voyages in sailing vessels, but I am informed that on the lee side of the large waves the vessel not only loses the wind but that the sails are sometimes actually "taken aback".

(4) The height of ocean waves measured from ships on their course

The following observations relate to the measurement of the height of waves from trough to crest in the open sea where they have room to grow to the full size which the wind is capable of producing, in water of great depth where the bottom can exercise no influence, and where there is no appreciable current.

I take first the region of the Trades, where the wind is continuous and not circulatory, maintaining a constant direction within one or two points of the compass.

In January 1907, between Puerto Colombia and Colon, the Trade Wind which followed us had at least the force of a Strong Breeze (number 6 of Beaufort's scale). The deck of the S.S. *Jamaican* was 16 feet above the water-line, and my eye-height as I sat in my deck chair 19 feet. From this point of view the wave-crests frequently topped the horizon.

At Colon, part of the cargo was unloaded, lightening the ship so that the height of the deck above the water-line was raised to 18 feet, and my eye-height, sitting in a deck chair, to 21 feet. On our new course we had the wind in our teeth. Its force was between a Strong

Breeze and Moderate Gale, with a velocity, therefore, of about 31 miles per hour, and the ship pitched in a heavy sea. The wave-crests were commonly just below the horizon, but occasionally touched it, thus reaching a height of 21 feet above the water-line. The length of wave which is produced by wind of the force then blowing is about equal to the length of the vessel in which I was sailing, and the waves were head-on. The trough of the wave must consequently have been somewhat lower than the floatation line in still water, so that the true height of the waves from trough to crest must have been somewhat more than 21 feet.

On this and other occasions I have been able to ascertain the height frequently reached by waves, but never an average which would include those during the smoother intervals, for when on a big ship it is not possible to find a point of sight as low as their crests.

The next observations relate to the height of waves in a Strong Gale (Force 9, velocity 50 miles per hour) in the North Atlantic. During a stormy voyage from Liverpool to Boston, Mass., in December 1900, on the Cunard S.S. *Ivernia*, there was one day, December 7th, during which the waves ran true before a Strong Gale, with no crossing sea and no visible swell. The position was Lat. 50° 56′ N., Long. 25° 33′ W., in very deep water, far from land, and away from the influence of currents. Six or seven long, steep-fronted ridges could be seen simultaneously, advancing to meet the ship on the port side at an angle of about 75 degrees with her course. The heights of the lower and upper deck were favourable for measurements, the waves commonly topping the horizon when I stood on the lower deck, and one of exceptional size topping the horizon when I stood on the upper deck. The vessel pitched but did not roll, so that by taking my station amidships the height of the view-point was constant. The force of the wind heeled the ship to starboard and held her so. This angle was measured and due allowance made. The height of the decks above the water-line was obtained from the scale-plan of the ship.

The resulting measurements showed that the ordinary waves rose more than 29 feet above the water-line and one above 43 feet. Once again during the voyage when the wind rose to a Strong Gale an exceptional wave rose above the 43-foot level.

An indication of the wave-length associated with these heights was obtained on December 8th, when the wind had abated and also changed direction, and a swell on the port side was evidently, from its size and direction, the continuance of the "sea" which prevailed on the 7th. The ship had now altered course, and the swell was nearly at right angles, but from a point slightly ahead. The interval at which the waves met the ship was 13 seconds, so that the true period was somewhat greater, from which it follows that the speed of the swell was more than 45 miles an hour, and its wave-length more than 866 feet. The length of the *Ivernia* was 600 feet, the true wave-length on December 7th probably about 900 feet, and the angle at which the waves met the ship was such that the distance between crests measured in the

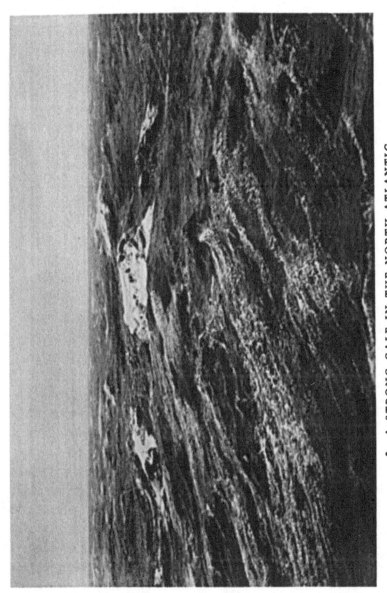

3. A STRONG GALE IN THE NORTH ATLANTIC

direction of the ship's course was several times the length of the ship, so that the vessel would settle to her proper floating line in the interval between. Thus the recorded height of the waves above the horizon was the full height from trough to crest.

On February 9th, 1907, I had another opportunity of measuring the height of waves produced by a Strong Gale (Force 9) in the North Atlantic. We were homeward bound from New York for Southampton on the Atlantic Transport Company's S.S. *Minnehaha*, in Lat. 48° 54' N., Long. 18° 20' W. Only a moderate northwest breeze was blowing, but a huge, steep swell overtook us from this direction (about 45 degrees abaft the beam), and the vessel rolled with a slow and perfectly regular motion.

A wireless message came from the S.S. *Cedric*, not far to the north of us on the outward course to New York, saying that the ship was in a Strong Gale from the north-west.

Standing on the lee side of the promenade deck of the *Minnehaha*, I watched the long-crested ridges as they passed away, and each in succession obscured a large arc of the horizon. An experiment in raising my point of sight by 2 feet left the horizon still obscured, so that I can safely add this amount to the eye-elevation in calculating the true height of the waves.

I measured the height of the deck above the water by means of a heavy rope which I hung over the lee side, and found that it varied from 29½ to 33½ feet with the oscillation of the water against the ship's side, an average height of 31½ feet. That this was correct is indicated by a measurement which I made two days later in the smooth water of the Solent, when, with rather lighter bunkers, the deck height was 32½ feet.

As each great swell passed to the lee, the deck heeled up, so that the ship was not on an even keel at the moment of observation. Fortunately the movement was quite regular and rhythmical, so that I was able to measure the additional height due to the roll, which was 2 feet. Thus adding 5½ feet for the height of my eye above deck when standing, I found that the height of the remarkably regular billows which obscured the horizon was at least 41 feet, a sum made up as follows:

Height of deck above water-line on even keel	31·5 feet
Height added by rolling	2·0 ,,
Height of eye above deck	5·5 ,,
Height of wave above line of sight (at least)	2·0 ,,
	41·0 feet

Taking account of the angle at which the waves overtook the ship and the length proper to the waves produced by a Strong Gale, it is evident that the ship would settle to her true floatation line in the long valleys between the ridges, so that the above measurement represents the full height from trough to crest.

Of the height of waves during exceptional weather, exceeding a Strong Gale (Force 9) but not of "Hurricane Force" (number 12 of Beaufort's scale), I can cite with confidence the observations of my friends Captain J. G. K. Cheret and the late Captain Percy Howe, not merely on account of sea experience, which is common to many published accounts, but because by means of question and answer I was able to assure myself that the circumstances were favourable to accuracy and that the measurements were properly comparable with those of which particulars have been given above. Captain Cheret's observations were made between Southampton and the Azores, Captain Percy Howe's between the Cape of Good Hope and Adelaide. In both cases the eye-height on the navigating bridge was 45 feet, and in the storms referred to, representing the extreme of these officers' experience at sea, the *ordinary* waves obscured the horizon.

To these records I will add the reliable observations of a Civil Engineer, my friend Mr G. T. Ogilvie, of which full particulars were supplied to me. On a voyage home from New Zealand in a thirteen-hundred ton ship with a length of 230 feet Mr Ogilvie made systematic observations, climbing the mizzen rigging to such a height as would enable him to sight-on to the horizon. Near Cape Horn, during a Full Gale from the south-west, waves of 30 feet from trough to crest were comparatively common. A good observation was made that day of one wave of 42 feet, and in the course of the voyage there were one or two waves which probably reached 45 feet. In a solitary case the data indicated 48 feet, but this was so much above the line of sight that Mr Ogilvie considered that the figure could hardly be reckoned as an actual measurement.

The height of waves as reported in the newspapers on arrival of an Atlantic liner after an exceptionally stormy voyage sometimes relates to the deluge which occurs when a vessel drives her bow into a head sea, not to that of the undulations running on the ocean. If the water comes over the roof of the navigating bridge and this be 100 feet above the water-line, we receive the report of a wave 100 feet in height. This is probably broken water, but even if such be called a wave the height is probably much less than stated, for the occurrence usually happens when the bows are buried in the sea and the bridge, which is a long way ahead of midships, is dipped far below the proper level.

I have, however, had the opportunity of going thoroughly into the circumstances of observation in two cases of wave observation during those very rare occasions when the wind in the 'forties of latitude in the North Pacific and North Atlantic Oceans reaches what is known as "Hurricane Force" (number 12 of Beaufort's scale). These occurrences must not be confused with the Typhoon or true Hurricane, which, although larger than an ordinary Whirlwind, is smaller than an ordinary Atlantic Cyclone. Both the storms to which I shall refer were of the ordinary cyclonic type of the temperate latitudes, but exceeded their customary scale in duration and area as well as in force of wind.

The first of these occurrences was narrated to me by Captain Wilson, Commander of the S.S. *Ascanius*, on which I was returning from South Africa.

In October 1921, when Captain Wilson was in command of a Blue Funnel steamer of 12,000 tons on the route from Yokohama to Puget Sound, a storm of hurricane force came on, and the barometer fell so low that the needle went off the barograph. The storm had a wide area and lasted a long time. The elevation of the observer's eye on the bridge was 60 feet. The bar of the awning framework at the end of the bridge was considerably higher, yet when the ship was on an even keel this bar was on a line with the tops of the waves. From this observation Captain Wilson was certain that the waves were as much as 70 feet in height and probably more.

The second record is from the S.S. *Majestic*. The following report by the Commander, Commodore Sir Bertram Hayes, K.C.M.G., and Lieutenant A. F. Butcher, Fifth Officer, was published in *Lloyd's List* for February 20th, 1923. (The position of the vessel at 10.30 p.m. on December 29th, 1922, was Lat. 48° 30′ N., Long. 21° 5′ W., in the right-hand rear quadrant of a deep atmospheric depression.)

4 p.m. Barometer steadily falling. Whole gale with very heavy squalls. Wind west. Between this time and 8 p.m. the weather grew gradually worse and more threatening; the rain squalls became more violent and at frequent intervals. A very high head sea and swell was running. From about 8 a.m. the ship was practically hove to—having only just steerage way.

8 p.m. Wind storm force with worse conditions than above.

8.30 p.m. Wind increased to hurricane force. The barometer reached its lowest reading at 9 p.m., 28·30. The seas were mountainous—estimated by observers to be about 80 ft. Rain squalls were constant and fierce; the wind blowing steadily from west to west by north.

10.30 p.m. The wind shifted to N.N.W., accompanied by a heavy rain squall. The hurricane force of the wind kept steadily up until about 12.30 a.m., when it showed signs of abating a little. The height of the seas at midnight was apparently phenomenal; experienced officers agreed that they had never seen such precipitous seas.

4 a.m. Wind had veered to No. 9, this continued up to noon, the force of the wind gradually decreased to 6.

Further notes were supplied by Lieutenant J. A. Heenan, R.N.R., Fourth Officer.

Having been invited by the Editor to discuss the observations, I asked for a scale-plan of the ship, which was supplied me; and Lieutenant Heenan also sent additional details of the circumstances for my information.

The observations made at about 10 p.m. from the bridge, where the elevation of the eye was 89 feet above the water-line when the vessel was on an even keel, may I think be fairly summarised as follows. When the bow dipped, and even when the vessel was on an even keel, the nearer waves appeared to rise well above the general level of the distant horizon, which however did not show up very clearly in the moonlight. There was, however, sufficient light to note the height of the approaching waves in relation to the crow's-nest, and it was noted that when the bow dipped the crests of these waves ranged from the top to the bottom of the crow's-nest. Its mean height is 11 feet above the line of sight from the bridge when the vessel is on an even keel. The fact that the propeller never came out of the water fixes the limit of the angle of dip, and shows that the bridge was lowered at most about 30 feet and the crow's-nest 40 feet. This, therefore, would give about 60 feet as the minimum measured height of waves. I draw the general conclusion that the observations are best satisfied by a range from 60 to 90 feet, which gives a mean value of 75 feet. The following table summarises the relation of the heights of ocean waves above recorded to the speed of the wind by which they were produced. The figures relate to the general size of the big waves which dominate the scene, not to the height to which a solitary peak of water, caused by over-riding, may shoot up, nor to the conditions during comparatively smooth intervals.

RATIO OF HEIGHT OF THE LARGER, RECURRENT
WAVES TO THE SPEED OF THE WIND

Beaufort number	Speed of wind in statute miles per hour	Height of waves in feet	Ratio of height of wave to speed of wind
6–7	31	21	0·7
9	50	35	0·7
10–11	63·5	45	0·7
12	Above 75	About 70	—

The height in feet of the larger recurrent waves during winds ranging from Strong Breeze to Whole Gale is thus 0·7 of the speed of the wind in statute miles per hour. The speed of the wind corresponding to Force 12 is merely stated in Form 317 of the Meteorological Office as above 75 miles per hour. If the above relation holds for exceptional storms in the temperate latitudes, the velocity of wind required to produce such waves as those recorded by Captain Wilson on the North Pacific and the officers of the *Majestic* on the North Atlantic would be 100 miles per hour.

A considerable difference in the height of successive billows on the ocean is not necessarily due to physical irregularity, being a normal result of combination of wind-wave and swell.

This is illustrated by a record of the late Mr Ralph Abercromby[1] in the South Pacific, between New Zealand and Cape Horn. This observer used a 4½-inch aneroid barometer with a very open scale divided in hundredths of an inch.

Taking one-thousandth of an inch as equivalent to a difference of one foot in level, he recorded differences of level from trough to crest of individual billows varying from 21 to 26 feet, but the absolute difference between the lowest and highest level of the aneroid was 35 feet. A complete series of these measurements would give the true height of both wind-waves and the swell.

(5) On the sea-room required for the full development of waves

We have next to look for evidence as to the distance from the windward shore at which full-sized ocean waves are recorded in the Atlantic. Reliable determinations were made by Dr William Scoresby in Lat. 51° N., Long. 38° 50′ W., with wind from west-south-west. Reckoning from the coast of Newfoundland, the length

[1] "Observations on the Height, Length and Velocity of Ocean Waves", *Phil. Mag.* Vol. xxv, 5th Series, April 1888.

of fetch (in the afternoon when the observations were made) was about 600 geographical miles. In this "Hard Gale" quite half the waves exceeded 30 feet; and 40 feet was not uncommon. There were also occasional peaks of water due to over-riding which may have reached 50 feet.

The meridian 40° W. is associated with full-grown storm-waves from a westerly quarter. Here the length of fetch from the American shore may be anything from 600 to 900 geographical miles, according to the direction of the wind. Thomas Stevenson's empiric formula for the relation of maximum height of wave to length of fetch, namely,

$$\text{Height} = 1 \cdot 5 \times \text{square root of length of fetch in}$$
$$\text{nautical (or geographical) miles,}$$

was based upon observations in localities where the length of fetch did not exceed 165 miles. If, however, it be applied to distances of 600 and 900 nautical miles the heights would be 37 and 45 feet respectively, which are in accordance with observations in the Atlantic.

It is, however, important to note that my observations from the S.S. *Ivernia* and the S.S. *Minnehaha* indicate that there is no further development of height in the next twenty degrees of longitude east of 40°. Thus it seems that when we pass the distance of 600 or 900 miles from the windward shore in the stormy regions of about forty to fifty-five degrees of latitude in the North or South Atlantic, we should no longer visualise a continuous straight-line run of wind and waves but

27

think in terms of travelling cyclonic depressions. It is evident that the position most favourable to the formation of large waves will be that part of the disturbance where the direction of the wind is the same as that in which the depression is advancing; for only along this line can the wind continually nurse the waves which it produces. Here, moreover, at about half-way between the centre and the fringe of the cyclone, the strongest wind is developed.

Meteorological records do not show any relation between the rate of advance of a cyclone and the velocity of wind produced therein, so that it is apparently a matter of chance whether the point of application of the wind advances at the speed most favourable for increasing the swiftest waves which this particular force of wind is capable of producing. As the arrival of a long, swift swell is a common warning to ships at sea of the approach of a severe storm, it is evident that the ordinary rate of a cyclone is less than the velocity of the waves produced by a Whole Gale. As regards the North Atlantic, this is confirmed by statistics, for the late Mr F. J. Brodie found that of two hundred and sixty-four gales advancing from the Atlantic upon the shores of Britain only sixty, or less than one in four, travelled at more than 35 statute miles per hour.

In deep water, however, the gravitational progression of a group of waves has only half the speed of the individual waves, and new waves are continually forming in the rear. It seems, therefore, that any rate of advance of cyclone between the speed of the wind and

28

half the speed would be consistent with reinforcement of the waves. The formation of the largest waves, however, is not necessarily dependent upon the continuance of such a combination of circumstances, for the strongest gales commonly come in groups, one succeeding another after a short interval of time. Thus there may be a stormy month during which one cyclonic storm quickly succeeds another, all pursuing the same general track across the ocean. Between times the sea never settles down but heaves with a heavy swell almost constant in direction, and having the speed of the swiftest waves which the winds can produce. No sooner does a cyclone brew upon the North Atlantic in such a season than the wind in the right-hand, rear quadrant of the depression travelling towards Europe immediately steepens this swell into great storm-waves, as happened in the Bay of Biscay on December 21st, 1911, during a storm of which I have already given an account.

We have seen that 600 to 900 geographical miles appears to be a sufficient distance from the shore for the development of the largest waves which can be produced by a Whole Gale. Even the exceptional waves recorded from the S.S. *Majestic* for wind of hurricane force were in a situation where the distance from the windward shore was not more than about 1200 miles, taking account of the direction of the wind at the time.

It is evident, however, that the broader the ocean the more frequent will be the opportunity for winds of maximum force to produce their full effect of wave-making. The opportunity is usually given by the presence of a swell of suitable speed and direction at the point where the storm originates.

Let us suppose that the swiftest possible waves have been produced in a run of 900 miles from the windward shore, and that the whole width of the ocean is only 1000 miles. It is evident that the resulting swell will have a very short life, for in another 100-mile run it will be destroyed by breaking on the beach. If, on the other hand, the ocean be infinite in breadth, a condition nearly satisfied in the Southern Ocean, where the westerly swell rolls round and round the world, the shorter waves flatten out completely and on the other hand the long swift waves enjoy the longer life which their greater store of energy makes possible. Great wave-length and great crest-length, rather than excessive height, are the characters which commonly attract attention in the Southern Ocean, and this is attributable in part to the fact that here the circulation of the wind is not broken up by land.

In the course of voyages on the North Atlantic I have had several opportunities of measuring the distances to which a swell will run without flattening out to invisibility. Thus on January 8th, 1914, on the voyage from Southampton to Colon, in Lat. 24° 26' N., Long. 43° 6' W., we encountered a swell from north of west travelling at 46 miles per hour, and on the following day a swell travelling at 40 miles per hour, and their directions traced backwards on the chart radiated from a point 1500 geographical miles distant.

On February 22nd, 1914, on the return voyage, in

Lat. 25° 14' N., Long. 44° 14' W., a low, swift swell from a little west of north invaded the Trade-Wind area. On the two following days this characteristic swell of stormier latitudes was higher, and its speed was measurable, 46·4 miles per hour on the 23rd and 45 miles per hour on the 24th. Tracing backwards the courses of the swell on the 22nd and 24th I found that they radiated from a point 1400 geographical miles distant, in the vicinity of "fifty north by forty west", a region whose stormy character has been described by Mr Kipling.

On February 28th, 1914, in Lat. 45° 17' N., Long. 11° 23' W., and on the following morning, March 1st, not far south-west of Ushant, the swell was about 18 feet high and had a speed of 42·7 miles per hour. The directions on these two days radiated from a point about 840 geographical miles to the west. These examples show to what great distances the waves characteristic of a Strong Gale, Force 9, will travel beyond the area of the wind which produced them.

The interesting question arises as to what is implied by the observed radiation from a point. Light is thrown upon this by observations which I made in the course of a voyage from Liverpool to Cape Town by the S.S. *Nestor* in 1929. On July 2nd in Lat. 40° 4' N., Long. 11° 46' W., a moderate westerly swell was recorded which crossed the ship's course during the two following days, its general speed being about 30 miles per hour. The position on July 4th was Lat. 29° 20' N., Long. 14° 59' W. Tracing backwards the directions of the swell observed on the three days, the six points of intersection lie close together not far from the Azores. The distance of their mean position from the nearest point of the ship's course was about 400 nautical miles. The important feature of this record is, however, the great angle of divergence between the swell on the first and third days, which was no less than 78 degrees of arc. This, I think, shows that the diverging swells were the result of waves simultaneously generated by the divergent winds in different parts of a cyclonic depression, not the result of lateral radiation from a single group of waves.

(6) *The steepness of waves on oceans, inland seas and lakes*

The following table generalises the results of observations recorded in the preceding pages relating to the maximum dimensions of the recurrent waves formed by winds of different speeds upon the open ocean far from the windward shore.

MAXIMUM DIMENSIONS OF THE RECURRENT WAVES
OF THE OCEAN IN RELATION TO SPEED OF WIND

Beaufort number	Speed of wind statute miles per hour	Speed of wave statute miles per hour $=\frac{8}{10}$ speed of wind	Period (secs.) =speed of wave÷3½	Length (feet)= square of period ×5⅛	Height (feet) $=\frac{7}{10}$ speed of wind	Length/ Height =period ×1⅔
6½	31	24·8	7·0	251	21·7	11·6
7	35	28·0	8·0	328	24·5	13·3
8	42	33·6	9·6	472	29·4	16·0
9	50	40·0	11·4	666	35·0	19·0
10	59	47·2	13·5	934	41·3	22·6
11	68	54·4	15·5	1231	47·6	25·9

The above table shows how much slighter is the slope of the final waves of a great storm on the ocean than that in a Moderate Gale or Strong Breeze. Taking account of the immense length from crest to crest of the principal waves in a Whole Gale or a storm of hurricane force, it seems likely that beyond the shelter of the eddy to leeward of each crest there would be sufficient water exposed to the direct current of the wind for the raising of steep waves some feet in height which would have a considerable effect in masking the longer waves.

Having raised this point as a suggestion for further observation, I proceed to consider the steepening of waves as we go backwards in the scale of length. The following records of the size of waves on the Western Mediterranean, Lake Superior and smaller lakes, prove that the steepening shown in the above table is not continued much further than the gradient of the waves produced in the open ocean by a Strong Breeze, where, as in the Trades, no crossing swell interferes with the action of the wind.

On April 7th, 1899, on board the Orient liner *Orizaba*, on the course from Marseilles to Naples, a Moderate Gale, Force 7, 35 statute miles per hour, blew directly following our course as we neared the Straits of Bonifacio. Taking a position amidships on the lower deck, I found that the crests of the waves topped the horizon when my point of sight was 22 feet above the floatation line. I had not then discovered the method of determining the period of waves by the rise and fall of foamspots (which I shall describe later) but I obtained a rough measurement of the period by means of a log of wood thrown overboard, and found it to be 8·3 seconds, corresponding to a wave-speed of 29 miles per hour and a wave-length of 353 feet. The ratio of length to height was, therefore, not more than 16. Thus the wave produced by a Moderate Gale in this part of the Western Mediterranean differed little from that in the Caribbean. The greatest possible length of fetch for these Mediterranean waves, having regard to the direction of the wind, was 214 nautical miles. If Thomas Stevenson's formula for the highest waves be extended to this distance, the figure would be 21·9 feet. It seems, therefore, that the length of fetch was sufficient for the unfettered action of a Moderate Gale. When, however, we enquire into the records of severe storms in the Western Mediterranean there are indications of insufficient room for full development. Thus in the Gulf of Lions the highest waves are reputed to reach 30 feet[1] in storms of a violence which in the North Atlantic is commonly associated with a wave-height of 40 feet. In the Gulf of Lions, therefore, with a possible length of fetch of 450 nautical miles, a Whole Gale, Force 10, appears to produce such waves as are raised by a Fresh Gale, Force 8, 42 miles per hour, on the open ocean. If Thomas Stevenson's formula were extended to include a fetch of 450 miles it would give the maximum height of the waves here as 32 feet, which seems to concord with observation.

On Lake Superior, waves are recorded of about the length and height which I have observed during a Moderate Gale in the Caribbean and Western Mediter-

[1] See *The Mediterranean*, by Admiral W. H. Smyth.

34

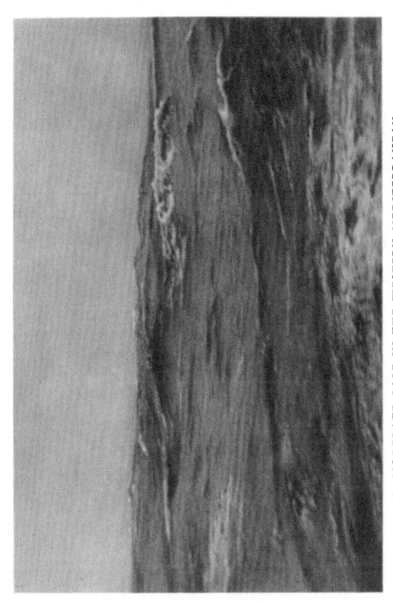

4. MODERATE GALE IN THE WESTERN MEDITERRANEAN

ranean, but they only occur on the lake during storms of exceptional severity such as happen only once in several years, in which I presume that the wind would not be less than Force 10 of Beaufort's scale, with a speed of 59 miles per hour. Particulars of these waves were given by my friend the late Colonel D. D. Gaillard, U.S.A., in his book on *Wave-Action in Relation to Engineering Structures* (Washington, 1904).

Ships' captains, who had navigated the lake for many years and had made observations of the obscuration of the horizon from the wheelhouse, reported that in storms of exceptional severity such as occurred only once in several years the height of the waves ranged from 20 to 25 feet and the length from 275 to 325 feet, corresponding on the average to a speed of 27 miles per hour.

On account of the frequency of error in estimation of wave-lengths, it is important to have some check upon the statement of length. This is fortunately provided by the observations made by Colonel Gaillard in 1901 and 1902 of the period of waves entering Duluth Canal at the west end of the lake. On the ten days of roughest weather the average period ranged from 7·4 to 8·25 seconds, corresponding to lengths in deep water of 281 and 349 feet, and an average speed of 27 miles per hour.

In January 1901 I recorded breakers with a period of 7·7 seconds on the north shore of the lake near its middle point. The corresponding length in deep water would be 304 feet. Thus we may safely estimate the steepness of the storm-waves from the average dimensions reported from the steamers, that is to say, height 22½ feet, length 300 feet; and therefore 13⅓ as ratio of length to height. But whereas waves of these dimensions are produced on the ocean and the Western Mediterranean by a wind of 35 miles per hour, the wind-velocity required for their formation on Lake Superior must be somewhere about 59 miles per hour. Thus while on the Atlantic and the Mediterranean such waves are subject to a wind-thrust of about 8 miles per hour, on Lake Superior, where the length of fetch would be at most 300 miles, they would experience a thrust of about 32 miles per hour. On the other hand, in a storm of such strength as to produce waves 22½ feet in height on Lake Superior, the height of the waves in the Western Mediterranean would be about 30 feet, and in the North Atlantic about 40 feet.

On the Lake of Geneva, which has a length of 45 statute miles measured on the curved central line, waves with a height of 8·2 feet are recorded in the data collected by Thomas Stevenson; and the late Dr F. A. Forel recorded a period of 5 seconds at the west end of the lake. The corresponding wave-length is 128 feet, so that the ratio of length to height, 15·6, indicates no further steepening.

The following observations on Coniston Water, with a length of 5½ statute miles, were reported to me by Mr Hamil, a former captain of the lake steamer. After three days of steady wind blowing up the lake had produced a steady run of waves, the wind came on to blow with unusual violence in the same direction, forming waves of such remarkable size that Mr Hamil set himself to measure their length and height against the side of his little vessel, then lying at the head of the lake. He found the length to be 65 feet, a measurement upon which he could rely within 5 feet either way. The height he determined as 5 feet, but found this measurement more difficult. The ratio of length to height is 13, which is the same as the steepness of the storm-waves on Lake Superior and that in a Moderate Gale on the Caribbean.

I come finally to a record relating to a sheet of water only one furlong in width, the Round Pond in Kensington Gardens. On a day when a wind-speed of 50 miles an hour was registered in London, I determined the period of the waves at the lee side of the pond, and found it to be one second. This corresponds to a length of 61·5 inches. A ratio of 13 would correspond to a height of rather less than 5 inches and the appearance of the waves was consistent with such a value. The little ridges were travelling at about 3·5 miles an hour, so that they were exposed to a wind-thrust of 46·5 miles an hour. If the wind had power to force the waves to greater steepness yet, as I suppose, the thinner ridges so produced would have sheared under the lateral thrust.

(7) *The reaction of ocean swell upon the wind*

I learnt from the experiences when hove-to in a storm, first that the estimate of wave-length by eye is subject to systematic error, and secondly that the theoretical wave-length for a slight swell is applicable to the conditions of a Strong Gale within the limits of observational error. The results obtained also suggested to me that the action of the wind when sufficiently prolonged was to produce waves travelling at nearly its own speed, an harmonious procession, in which the waves were no longer buffeted, but fanned by a gentle breeze. In my investigation of this idea, I was hampered at first by the lack of a convenient method for determining the period and the speed of waves during a voyage. The direct determination of velocity by timing the passage along the ship's side is only possible with waves of considerable size. Moreover, I have never been able to get any sure result from the promenade deck of a liner except by enlisting the co-operation of a second observer. Co-operation is useless unless continued until concordant results are obtained, and such prolonged co-operation is not easy to obtain. However, during a voyage from Southampton to Colon and back in 1912, I found a quick and simple method for determining the period of the waves which enabled me without assistance to ac-cumulate rapidly a great number of measurements, a matter of prime importance in phenomena so variable. The method consists in observing the rise of a spot of spent foam to the crest of a wave, starting the stop-watch and, having waited for the foam-spot to descend to the hollow and rise on the next wave, stopping the watch when the foam again reaches the crest. The observation is then repeated and the average of ten such periods taken. They are not of consecutive waves, but the whole set of measurements is obtained in the course of a few minutes. A second set of ten is then determined, and it has always been found that the two means are closely concordant. A great advantage of the method is that it is applicable both to small and large waves, and that it enables the period of both the waves and swell to be determined, even when, as is common when the Trade Winds are blowing moderately, the sea has an extremely confused appearance owing to the over-running of the waves by a concurrent swell not very different in speed. Thus two series of ten intervals of the larger undulations were timed, one directly after the other. The average of the first was 7·97 seconds, of the second 8·03 seconds. The minor inequalities were similarly timed and gave periods of 6·09 and 6·24 seconds.

The fact that foam is carried along by a tidal or other current contributes to accuracy, for it enables the true interval between waves to be determined without measuring the current; but if the wind have a purchase on the foam so as to make it drift faster than the water some error will be introduced. When I first used the method of determining the period of waves from the foam-spots I had already studied the action of wind upon a frothy surface, and had photographed its rippling effect. I noticed particularly that the white patches which remain for a long time on the sea are mere films, and that the wind does not deform them. It was evident, therefore, that if there were any drifting by the wind it must be slow. At length I was so fortunate as to obtain good conditions for measuring the amount. The place of observation was the ferry pier at Burton-upon-Stather in Lincolnshire, near the outfall of the Trent into the Humber. A fresh breeze, the speed of which I estimated at more than 20 miles per hour, was blowing straight down the river, making short, steep breaking waves, and there was much foam. At the beginning of the observations the tide was ebbing, so that wind, wave and tide were all concurrent. Then the tide turned, and the flecks of foam travelled with the flood tide through the little, angry, breaking waves and against the "Fresh Breeze", which in the countryside we call a "Strong Wind". I collected pieces of wood, as dense as I could get, so that they settled deeply in the water and exposed little surface to the wind, and measured the rate of flow of foam-spots and drift-wood, both before and against the wind during the ebb and the flow of the tide. The result was that the foam-spots showed a drift due to wind and breaking wave acting together of half-a-foot per second. Even if this were wholly due to the drag of the wind it would be small in relation to the speed of ordinary ocean waves, but a considerable part of the drift appeared to be due to the rapid succession of buffets from the short, steep waves. Foam drifting half-a-foot per second would travel 5 feet in 10 seconds, whereas the crest of a 10-second wave advances 512 feet in that time, and an error of 1 per cent. is of no account, having regard to the degree of irregularity in the waves themselves.

Having now a quick and accurate method of measuring the period of waves at sea, and therefore of calculating their speed and length, I desired to measure the velocity of the wind at sea by instrumental means, since the esti-

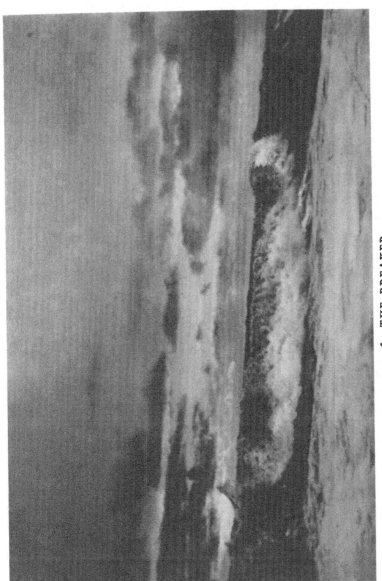

5. THE BREAKER
(Eastbourne)

mates of wind-velocity in Beaufort's scale were not in the same class of accuracy as my determinations of the speed of waves. Indeed, I found that for light winds the entries in a steamer's log are not as accurate as would be the case if the officer on the watch could observe the action of the wind upon sails, which was the observational basis of Admiral Beaufort's elaborate scale.

Sir Napier Shaw, F.R.S., then Director of the Meteorological Office, having kindly lent me a Robinson anemometer (the instrument with four arms revolving horizontally provided with cups at the end), I embarked at Southampton on the S.S. *Oruba* of the Royal Mail line for the return voyage to Colon via Barbados and Trinidad. On the outward voyage the anemometer was placed aft on the Marconi house, where it frequently lost wind. On reaching Trinidad on the homeward voyage the anemometer was transferred to the navigating bridge and mounted at the starboard end, and here it was fully exposed, the wind continuing on the starboard quarter from February 18th up to and including the morning of the 28th. As a counsel of perfection it would be best to have an anemometer at each end of the bridge so as to avoid liability to sheltering by the hull of the vessel when the resultant draft of air is from the lee. The table of observations given on a subsequent page does not, however, include any case where the full exposure of the anemometer was in doubt.

The outward voyage had been so far profitable that I had become expert in the measurements required, namely the determination of the direction of the smoke, which shows the resultant current of air due to speed of wind and speed of vessel, and the determination of the direction of the waves. I embodied each day's measurements on the page of a large sketch book so as to have the results visibly displayed. One line showed the course and speed of the ship, a second, drawn from one end of this line, showed the direction of the smoke-track, and a distance was marked upon it equal to the velocity registered by the anemometer. A third line joining this point to the extremity of the first line completed the triangle and gave the true direction and true speed of the wind.[1]

In the course of the eleven-day voyage between Trinidad and the entrance of the Channel there were two days of variable airs on the northern border of the Trade-Wind area, but on the other nine days the wind was steady in force and direction, and as I had mastered the methods of observation I was sure that the measurements were of sufficient accuracy.

As so often happens in original research the results were quite contrary to expectation. I had anticipated

[1] A practical point affecting meteorological records emerged from the observations made on the voyage out and back. It is not unusual to note the direction of the "curl on the waves", that is to say their direction of breaking, as a means of deciding the true direction of the wind from a vessel under way. The direction of breaking is more definite and more easily recorded than the general run of the waves. I found, however, that whereas the direction of the general run of the waves coincided with the direction of the wind inferred from the anemometer reading and the other data, that given by the breaking water was seriously out of accord whenever the swell crossed the waves at a considerable angle. The direction of breaking is, in fact, the resultant of wave and swell, so that the general run of the waves provides a more reliable index of the direction of the wind.

that the speed of wind and wave would be concordant, with only a small excess of the former, and perhaps a constant ratio. But although the wind only varied between 16·4 and 23·6 miles per hour the deficiency of the waves' speed varied from 1·2 to 7·9 miles per hour, and the ratio of the speed of wave and wind from 0·60 to 0·93. The measurements seemed to contradict the idea either of a constant difference or constant ratio, and the whole voyage would have yielded a merely negative result but for the fact that I had kept a careful record of attendant circumstances, more from conformity with the rules of scientific observation than from any prevision of their utility.

It was after my return home and when I settled down to the analysis of the data, that I found a single, determining cause for the great, and apparently capricious, variation between the speed of wind and wave on this fair-weather voyage. The disturbing factor was the swell. When swell and wave followed the same course there was very little difference between the speed of wind and wave. When the swell followed at an appreciable angle there was a marked difference, and when the swell met the waves, either directly or obliquely, there was a very great difference between speed of wind and wave. The whole matter is expressed in the accompanying table of figures more concisely than is possible in words.

On the two days when swell and wave were in the same sense and direction the former travelled over the latter at an average speed of 11·55 miles an hour. The average difference between speed of wave and wind was

SPEED OF WIND AND WAVE BETWEEN TRINIDAD AND USHANT

Date 1914	Lat. N.	Long. W.	Speed of wind (statute miles per hour)	Speed of wave (statute miles per hour)	Difference of speed	Ratio of speeds	Speed of Trade-Wind swell	Speed of Atlantic swell	Angle between wave and	
									Trade-Wind swell	Atlantic swell
Feb. 18	12° 0'	61° 0'	16·4	15·2	1·2	0·93	29·5	—	0°	—
19	15° 22'	57° 1'	23·6	21·1	2·5	0·89	29·9	—	0° (15 feet)	—
20	18° 33'	53° 0'	20·4	16·5	3·9	0·81	28·7	—	Less than 45°	—
21	21° 52'	48° 43'	21·9	18·6	3·3	0·85	29·9	—	Less than 45°	—
22	25° 14'	44° 14'	22·6	16·3	6·3	0·72	24·6	—	Less than 45°	95° (slight swell)
23	28° 38'	39° 18'	19·0	15·9	3·1	0·84	—	46·4	0°	95° (slight swell)
24	32° 2'	34° 13'	Wind variable	—	—	—	—	—	—	—
25	35° 17'	28° 50'	Ditto	—	—	—	—	—	—	—
26	38° 37'	23° 11'	16·4	10·4	6·0	0·63	—	39·3	—	At least 135° (moderate swell)
27	41° 54'	17° 21'	19·4	11·7	7·7	0·60	—	39·5	—	93° (swell higher, 12 feet)
28	45° 17'	11° 23'	20·2	12·3	7·9	0·61	—	47·4	—	82° (swell 24 feet in height)

then only 1·85 miles per hour and the ratio 0·9. On February 27th and 28th, when the swell crossed the wave-pattern squarely at an average rate of 43·45 miles per hour, the average difference of speed between wave and wind was 7·8 miles per hour and the ratio 0·6.

Another instructive comparison is provided by the observations of February 21st, 22nd and 23rd. On the 21st, when a Trade-Wind swell of moderate height overtook the waves slowly at an oblique angle, the difference of speed between wind and wave was 3·3 miles per hour and the ratio 0·85. On the 23rd there was a very slight Trade-Wind swell in the same direction as the waves, and a slight Atlantic swell crossed the waves squarely at a speed of 46·4 miles per hour, the difference of speed between wave and wind was 3·1 miles per hour and the ratio 0·84. On the 22nd the Trade-Wind swell crossed the waves obliquely from behind, and a slight Atlantic swell crossed them squarely. On this day the difference of speed between wave and wind was 6·3 miles per hour, which is almost exactly the sum of the two effects observed on the 21st and 23rd respectively, and the ratio between speed of wave and speed of wind dropped from 0·85 on the 21st to 0·72 on the 22nd.

I had paid particular attention to the breaking of the water when wave and swell crossed, and sought for some relation between the retardation of the waves and the violence and frequency of this destructive interference. I found, however, that the height of the swell had less influence than its relative speed, although the amount of broken water was greater with the high swells than with the low swells crossing the waves swiftly. I explored many avenues in the endeavour to find a numerical relation between the deficiency of wave-speed and the conditions of conflict between the waters of the wave and swell, but every avenue was a blind alley. This drove me from thinking in terms of the visible conflict of water with water (where the undulations are not infinitesimal) to picturing in my mind's eye the effect of the swell upon the flow of air across the disturbed surface of the sea. Here at last there appeared to be a reasonable explanation of the retardation of the waves, for although a low, swift Atlantic swell meeting them did not deliver a strong buffet, but passed through quietly, yet the swift passage of one corrugated surface through the other must have completely disarranged the sequence of aerial vortices and superposed curves of flowing air by which the waves of the sea are moulded, driven and developed to their greatest height and speed.

The wind-waves on this fair-weather voyage were generally too low to admit of measurement from the decks of a large liner, but it was evident from the general appearance that they were much flatter than the waves produced by the same force of wind in waters free from a crossing swell. In other words, the action of the swell was to reduce height in a greater proportion than length, and therefore in a vastly greater proportion than speed. Thus on February 26th, when a moderate Atlantic swell met the waves at a sharp angle, these attained a speed of 10·4 miles per hour under the action of a wind of 16·4 miles per hour, whereas if there had been no swell

46

6. WAVES AT LOW TIDE
(Eastbourne)

their speed as calculated from the table given in a previous section would have been $16.4 \times 0.8 = 13.1$ miles per hour. Thus they had lost one-fifth of their speed. The reduction in wave-length was therefore from 72 to 46 feet, or more than one-third. These waves appeared to me to be scarcely 1 foot in height, and the navigating officer on the bridge at the time (who on this and a former voyage had joined in my observations) concurred in the estimate. I am aware that the error in such an estimate may be considerable, but since the full heights of waves with lengths of 72 and 46 feet are about 6 and 4 feet respectively, I think it is safe to infer that the height was reduced in a greater proportion than the length. The shortness of the ridges was quite as remarkable as the insignificance of height, the sea presenting no distinct pattern of parallel crests. The experience was in sharp contrast to that in the Red Sea on a day when the wind was of similar strength. I was *en route* from London to Yokohama by the S.S. *Hitachi* of the Nippon Yusen Kwaisha in February 1903. The ship made 12 knots, that is 14 statute miles, per hour, and a northerly wind dead aft held steadily at exactly the speed of the ship, the smoke going straight up from the funnel with no ventilation anywhere on board and no relief from the heat. There was no swell, and the sea presented the appearance of a regular procession of steep waves with long crests, every member of the procession apparently several feet in height.

The following observations made during calms show how incompatible is the changing surface produced by

47

crossing swells with the maintenance of wind vortices with horizontal axis suitable for the development of regular waves. When near the Azores on the voyage from Southampton to Barbados there fell a dead calm. The surface of the sea was glassy, but a considerable swell from the north-west met another of equal size from the south-east. Their meeting crests coalesced in a great barrel-vaulted billow which was almost immediately furrowed along the top, and then each re-constituted swell travelled away in its original direction, undiminished and unchanged.

Again, during a voyage from Liverpool to Cape Town, on entering the calm of the Tropics, two sets of swell, from the north-east and south-east Trades respectively, crossed each other at right angles, and the whole of the great watery plain was parcelled out in spherical mounds and cup-shaped hollows, undulating continually with undiminished amplitude hour after hour. With these observations made during calms may be contrasted those of my friend the late Captain Percy Howe, a skilful observer of the movements of water, who informed me that he had noticed both in the South Atlantic and South Indian Oceans that, with wind from the north-west, the waves which it was driving began to diminish in size as soon as a south-west swell came along, although the force of the north-west wind was undiminished.

Applying these observations to the greatly different conditions of lakes and oceans, it is easy to understand why the quick rising of waves on lakes is the inevitable and invariable rule, whereas on oceans the development of waves is very slow except when the wind comes on to blow in the direction of the swell prevailing at the time, and with a speed greater than that of the swell. When, however, this exceptional condition occurs, the development of waves is even more rapid than when wind blows upon smooth water, as was the case in the storm of December 21st, 1911, in the Bay of Biscay, described at the beginning of this chapter.

CHAPTER II

WAVES IN SAND AND SNOW FORMED AND PROPELLED BY WIND AND CURRENT

(8) *Waves in sand formed and propelled by wind*

The situation of my former home on the cliff at Branksome Chine overlooking Bournemouth Bay not only impelled me to study the waves of the sea but, by the proximity of a sandy shore, provided opportunities for the observation of the action of wind to ripple the loose, dry sand above the level of the tide and to drive the transverse ridges in orderly procession as a group of waves.

I was out on the broad sandy beach between Poole Haven and Studland after a morning of heavy showers. Presently, as the sand dried it began to drift, though not very copiously, and to leeward of the low, grassy dunes this driven sand dropped upon the smooth, slightly moist surface of the beach. The deposit was quite smooth at first but when a layer of about one-eighth of an inch thick had accumulated (that is to say a depth of about 4 or 5 sand-grains) the surface assumed a mottled appearance, and this irregular inequality was quickly transformed into a pattern of transverse ridges of a desk-like shape, with gentle slope on the windward and steep slope on the lee side. These were $1\frac{1}{2}$ inches from crest to crest, and travelled in a group as true waves at the rate of about a quarter of an inch a minute during the squalls of wind, remaining stationary during the lulls.

On another day when I was walking among the sand-hills of the promontory which protects Poole Haven on the Bournemouth side of the entrance (at that date a lonely spot) the sand was drifting very freely, for a fresh breeze was blowing after a fortnight of dry weather. A tract of loose sand was in regular ridge and furrow at right angles to the wind, the crest-length several feet, the wave-length, or distance from ridge to ridge, about 3 inches, both at the rear and front of the group, there being no increase of wave-length to leeward as in the case of the waves on a pond.

I tried the effect of isolating a group of ten ridges by smoothing the surface of the loose sand in front and to the rear. Marking the position of the ridges by inserting stalks of the marram grass, I watched the advance of the front and the rear of the group, both of which moved at the same rate. The speed of advance of the ridges throughout the fresh breeze was about a quarter of an inch per minute, but most of the advance was accomplished by a quick, visible shift of the ridges during the stronger gusts.

Continuing the observation of the aeolian sand-ripples on this shore throughout the winter, I noted the following characteristics.

The crest-length increased in a much more rapid ratio than the wave-length. This process appeared to be accompanied by the obliteration of ridges of lesser height, the vestiges of which were noticeable at Y-shaped junctions with the higher ripples. Examining the sand closely, sometimes with the aid of a magnifying glass, I noticed that the troughs were scored in the direction of the wind, showing that the hollows were not places of stagnant air. I also noticed, on an occasion when sand was accumulating and at the same time being rippled by the wind, that the surface was not continuous but composed of scattered sand-grains which darted hither and thither as ants when the nest is disturbed.

One definite and important result was obtained from the observations on this foreshore, namely, that the aeolian ripples in the sand maintained a constant ratio of length to height (18 as a round number), their profile not changing with growth. The measurement of wave-length was a simple matter, that of the height not so easy. The best way was to stretch a thread above the ripples and measure downwards to crest and trough.

Another set of measurements was of particular interest, namely that of the number of layers of sand comprised in the ridge. Making a section with the blade of a knife, very gently, so as not to disturb the piled sand-grains, I lay face down upon the sand and counted the number of layers. The two following examples are typical of the results:

Aeolian sand-ripples (measurements in inches)

Wave-length	Height	Height/Length	Number of layers
$1\frac{3}{4}$	$\frac{1}{10}$	17·5	8
6	$\frac{5}{16}$	19·0	16

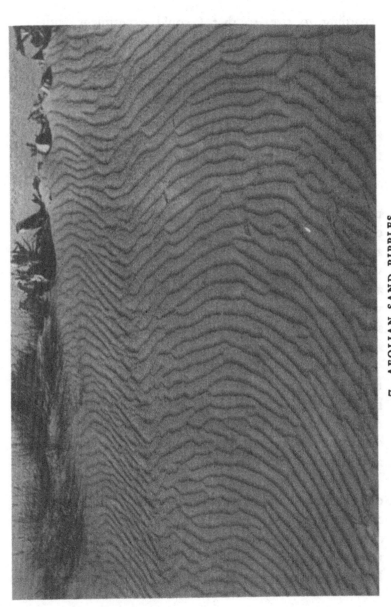

7. AEOLIAN SAND RIPPLES
(Southbourne)

according to size, so that I could experiment with materials of a uniform grain, as well as with natural sand.

I spread out a bed of the finest assorted sand and turned on a powerful blast. Almost immediately the sand fell into regular ripples of the usual profile, but arranged in curved ridges which were everywhere at right angles to the wind which radiated from the nozzle. Much of the fine sand flew up, and darkened the room with a choking haze.

The coarsest of the sand samples was of a kind which might be called small gravel, and in the terminology of desert deposits is known as "lag gravel". I spread out a bed of this, and turned on a moderate blast, before which it moved freely, but the surface remained level; and although under the action of a full blast the grains were driven along so swiftly and violently that they rattled loudly as they struck each other, yet there was no sign of rippling.

I then flung handfuls of the fine sand upon the coarser material and rippling occurred immediately. This would be expected, seeing that the fine sand by itself had rippled, but in the mixed material it was noticeable that the large grains of "lag gravel" got together transversely at the crests.

These experiments clearly indicated that aeolian ripples cannot be produced by mere rolling of sand, but have to do with lifting and suspension. Nevertheless the presence of sand-grains too heavy to lift helped the formation of ripples by accumulating in transverse barriers of inter-locked grains which created a scouring vortex on the lee.

In the fine sand of the foreshore, rippling began before I noticed any assortment. Having regard to the observation already recounted of the discontinuous character of the surface layer of sand when rippling was proceeding actively in a place of deposition, it seems likely that the initial resistance which started the rippling was the interlocking in transverse lines of sand-grains not much larger than the average.

In April 1899 I went to Egypt in order to study the action of wind upon sand under desert conditions. I stayed first at Ismailia, where there is a considerable accumulation of sand-dunes bordering on Lake Timsah and marshy tracts in its neighbourhood. At this season a moderate breeze blew from an easterly direction during the sunny hours of the day. Upon a large deep deposit of sand west of a marsh this wind had formed a sharp crest about 18 feet in height facing westerly. The somewhat rounded weather slope was rippled. On the lee was a smooth steep slope of slipping sand below which was another rippled slope in which the ripples faced towards the cliff, being formed by the return current of air. Further to leeward was a sort of rippling which I had not seen upon the seashore. This was of greater wave-length, much greater steepness, irregular front, and not desk-shaped, but with the crests near the centre. These crests were entirely composed of the coarsest sand or fine gravel. Here it was evident that depletion had proceeded for some time, and that the sorting of the sand-grains due to tossing away of the finer sizes had reached a stage where the lag gravel had begun to form a sort of plated surface, so that the individual grains could no longer roll freely. Thus a condition had been established in which the ridges no longer perfectly satisfied the state of *metabolic progression* which is one of the two chief criteria of surface waves, the other being the occurrence in groups, although the conception of a "solitary wave" is not excluded from mathematical theory.

The sand in the troughs of these ripples was fine but somewhat compacted, the erosive action of the wind having cut down to deeper layers. The length from crest to crest was 7 feet and the height more than 6 inches, the ratio of length to height being therefore 14 as compared with 18 for those in the loose sand of the seashore.

On a nearly flat bed of deep sand I measured the length and height of an uninterrupted series of thirty-seven ripples which were of the ordinary kind except that a few showed the beginning of the excavated and compacted character, and were evidently somewhat steeper than those on the weather slope of the dune. The average wave-length of the whole series was 23·7 inches, the height 1·43 inches and the ratio therefore 16·6, which is only a little steeper than the ripples measured on the shore of Bournemouth Bay. The beginning of erosion probably accounts for the difference. The general size of the sand of this dune, mainly composed of quartz, was about the same as that of the shore. Thus of a sample taken from the slipping cliff, where the sand is remarkably uniform, 94 per cent. was between $\frac{1}{48}$th and $\frac{1}{96}$th inch in diameter.

8. AEOLIAN SAND WAVES
(Helwan, Egypt)

At Helwan, where I stayed after leaving Ismailia, the rock does not weather into sand but dust, and if a handful of this be thrown into the air when the brisk desert breeze is blowing it floats away as a cloud instead of subsiding slantwise through the stream-lines of the wind after the manner of rocky particles comprised within the range of magnitude which we associate with the term "sand". But at the time of my visit the Nile was low, and on the further side of the river at the ferry to Sakhara a broad stretch of sand-bank was exposed, all of which above the water-level had been thoroughly dried by the desert wind. When I landed there I found this dry sand modelled in a great group of waves, some hundreds of yards from front to rear, moving before the wind. In the case of one ridge I recorded a movement of 30 inches in 48 hours. The average distance from crest to crest was about 30 feet.

It was evident that these were not the familiar aeolian sand-ripples grown to a large size, for their surface was covered with ripples a few inches in length from ridge to ridge, well marked by a coarser grain of sand along the crests.

Sand-waves of the order of magnitude of 30 feet from crest to crest had been absent at Ismailia, and it was evident that their formation was due to the greater mobility of the fine Nile sand which the ordinary day-time breeze was able to drift more freely than the quartz sand of the desert. This Nile sand was not rounded but splintery, and composed of various minerals derived from eruptive rocks. Of the sand at Ismailia only 4 per cent. passed through a mesh of $\frac{1}{70}$th inch, whereas almost all the Nile sand passed through this, and much of it through a mesh of one-quarter of the diameter. Where dry, the sand seemed quite loose; I never noticed any deposit which appeared to have become compacted or "set" as the result of pressure.

These aeolian waves resembled the ripples in loose sand in having a desk-shaped profile with long slope facing the wind, but differed in the fact that the crests were not level but undulating. The degree of this undulation differed on different days and in different situations on the sandy foreland, and increased with increase of wind. It was evident, therefore, that the measurement of the ratio of length to height of a series of waves would have no definite significance unless the series comprised many members. I therefore took measurements (in a straight line down wind) of a continuous series of twenty-three waves on May 8th and of twenty-four waves on May 10th on a neighbouring situation. The results were as follows:

Date	Average wave-length	Average height	Length/Height (calculated from these averages)
May 8th	30 ft. 6 in.	1 ft. 8 in.	18·3
May 10th	27 ft. 6 in.	1 ft. 6 in.	18·3

The circumstance that the average steepness appeared to be independent of any difference in the degree of undulation of the ridges on the two days points to the conclusion that the sand scoured from the saddles was piled up on the peaks. The fact that the crest-line of these sand-waves sometimes varied in height so greatly

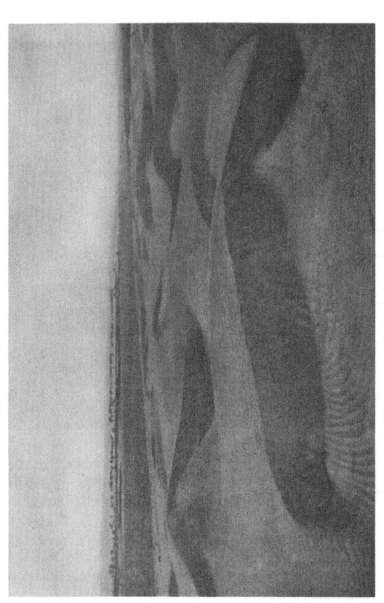

9. ABOLIAN SAND WAVES WITH UNDULATING CRESTS

(Helwan, Egypt)

as to have a definite peak-and-saddle structure evidently required a dynamic explanation, for the wind from day to day was constant in direction, so that the appearance was not due to furrowing by the action of a subsequent wind from another direction. The appearance of the ripples along the line of a saddle was significant, the sand on the crests here being distinctly coarser than on those of the ripples elsewhere, showing that the tossing action of the wind was here considerably increased. The current of air is, of course, deflected from right and left into this channel, and since the two sides are never exactly identical in slope and height the currents cannot be perfectly balanced, and so their junction must give rise to a spinning movement, not, as on the lee of a transverse ridge, around a horizontal axis, but revolving more in the manner of the travelling whirls which are made visible by water-spouts and "dust-devils". Such a mode of motion would result in a much greater increase in power of transporting material fine enough to be thrown into suspension than would be accounted for by any local increase in the horizontal velocity of the wind.

On the lee of the ridges the current of air converged from either hand towards the peaks (an action which was recorded by the direction of the ripples on either side), but the resulting effect upon the form of the sand deposit was not the same under all conditions. Sometimes the loose sand was scoured out of the trough so as to expose the underlying, compact bed of damp sand near the river level, leaving a nearly circular pit with a flat floor. This floor was bounded at front and back by the lee and weather slopes of consecutive sand-waves, and on either side by the loose sand pushed forwards from the saddles. Where the process had been long continued, the pits were sometimes the most conspicuous part of the pattern, the sand-sea having a pitted rather than a hummocky appearance.

But there were other conditions of sand-transport in which, instead of a scouring-out of pits behind the higher parts of the ridges, there was a piling-up of sand against the lee side of the peaks, in longitudinal structures which, viewed in combination with the transverse ridges, made a very striking pattern.

Which of these two very different patterns was produced evidently depended upon whether the charge of sand in each of the converging currents was such that the whirlwind at their confluence was overcharged with sand or capable of picking up more. In the former case, sand was thrown out of suspension and formed the longitudinal ridge along the line of meeting of the currents in the lee of the peak; in the latter case, the upward swirl where the currents met sucked up the loose sand in the lee of the peaks and, on account of the curved outline of the whirlwind, the sand was re-deposited in such a manner as to impart a nearly circular shape to the floor of the pit.

To leeward of the sea of loose sand was a bare stretch upon which sand was dropped from the cloud which filled the air up to a height of 20 or 30 feet. The loose sand collected in patches which at first had an oval form but soon became blunter on the lee side, higher along

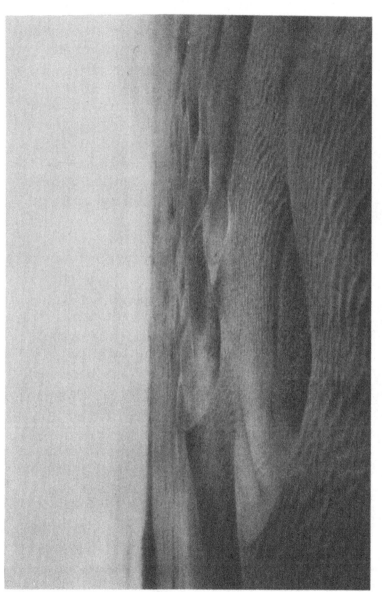

10. THE SAND PITS CALLED *FULǦES*

(Helwan, Egypt)

the central line, steeper on the lee, and with a cliff under the central part of the accumulation. At the same time the outline of the lee face became concave, the ends having the shape of the horns of the moon, and so was formed a miniature crescentic dune of the form which, on the large scale, is called a Barchan in the deserts of Africa and Central Asia, and in Peru a Medaño.

I inferred that similar forms would be left as residual structures if depletion continued where the sand-sea had assumed the pitted appearance, for it is evident that, if the sand on the lee side of the pits were removed, a collection of crescentic mounds would remain.

The crescentic Medaños of Peru, which travel before the wind and preserve their individuality by an equality of loss and gain of their component sand-grains, are a very interesting study. In respect of metabolism they resemble both waves and clouds, but it seems from the photographs that they dot the plain as the sky is flecked when cumulus is forming, and are not ranged in dependent sequence in the manner associated with a group of waves.[1]

Great pits in the deep sand-seas of Arabia, similar to those which I saw in the little sand-sea on the Nile foreland, are known as Fuljes, and the observations which I made are suggestive of their origin. When, however, desert sands are piled to a thickness of hundreds of feet it is possible that a setting or consolidation occurs in the lower layers, and I have not had the opportunity of ascertaining by observation whether the large pits are

[1] See *The Crescentic Dunes of Peru Appalachia*, Vol. XII, No. 1.

61

entirely due to the tossing away of loose material or partly formed by excavation of compact but friable material and the tossing away of the detached particles.

In the course of my visit to Egypt I camped out for a couple of nights among the great sand-dunes east of the Suez Canal on the route from El Kantara to El Arish. In the neighbourhood of Bir El Nuss these were commonly about 200 feet in height, long-crested, and arranged in a regular series. Judging from the detailed map made some years later, which is reproduced as Plate XVI in Vol. 1 of Dr W. F. Hume's *Geology of Egypt* (1925), these are enduring features of the landscape.

Their general profile was desk-shaped, but the crests were strongly undulating, with deep saddles, and here and there beautifully pointed peaks. The height from saddle to peak was of the same order of magnitude as the height of the saddle above the trough.

The details of form did not suggest that the peak-and-saddle structure was due to the crossing of one set of ridges by another subsequently formed by a later crossing wind, everything pointed rather to the explanation that the ridges had grown to a height at which the force of the wind was so much increased when drawing over the crest that the sand gave way altogether in weak spots so that furrows were produced, a peak-and-saddle structure resulting as in the rudimentary dunes of finer sand near Helwan. Sketches made by Sir Francis Younghusband, Dr Sven Hedin and others show how strongly the sharp peaks in a sand-sea impress the eye of the traveller, and

II. CRESCENTIC DUNE ("BARCHAN")

(Helwan, Egypt)

the interest of these striking features is certainly enhanced if, as I suggest, they are not necessarily the consequence of a change of wind but a form marking the attainment of a limiting size.

The measurement of a profile of a train of eight desert sand-dunes made by Dr Sven Hedin in the Takla Makan desert[1] is of special value in relation to the study of the wave-making action of wind exercised on the large scale in granular material, for the sand lies deep in the hollows and yet the form of the crests and troughs indicates that the outline impressed by the strong, creative winds has not been blurred by subsequent gentle breezes in which the air at the foot of the lee slope would remain almost stagnant.

The heights ranged from 100 to 200 feet, and the average ratio of length to height was 24·7. Another series of sixteen transverse sand-dunes near the Lower Tarim, of which the largest was more than 300 feet high, had an average ratio of length to height of more than 20, and the survey of those on the El Arish route in Northern Sinai also indicates a steepness less than that of aeolian ripples in quartz sand, and of the waves of a foot or two in height in the fine splintery sand of the foreland by the Nile.

These are but few measurements on which to base a generalisation, but they certainly suggest that the force of the wind is so much increased in passing across ridges

---

[1] Figured on p. 409, Vol. II, of *Scientific Results of a Journey in Central Asia*. This important figure has no number attached, but comes between those numbered 183 and 184.

of 100 feet or more in height that its power of erosion at the crests is too great for the sand to stand against it as steeply as in the smaller waves.

The measurement of angles of slope made by Dr Sven Hedin at different points on the profile of the very perfect group of eight dunes in the Takla Makan is of particular value. At, or very near the summit, is a sharp edge where the slope changes from a few degrees upwards to a dip of 30 degrees or more. Near the trough, the downward slope is gentle, and the beginning of the ascent is very gentle indeed, the steepest part of the weather slope being situated one-quarter of the wavelength from the crest. The form inevitably suggests that about three-quarters of the wave-length is modelled by the under-surface of a long, elliptical vortex, and that near the steepest point of the weather slope the direct current of the super-incumbent air strikes down upon the sand and drives it forward.

Unfortunately we have no record of observations during storms when the wind is operating on this scale in the Takla Makan, indeed it would probably be impossible to observe at all under such conditions. I recall that when the fine sand filled the air to a height of some 20 or 30 feet on the foreland of the Nile, I had to protect eyes, mouth, nose and ears, and even then suffered considerable discomfort during a moderate breeze. In drifting snow, fortunately, observations are easier, partly because one can swallow the snow with impunity.

One more point remains to be mentioned with reference to the conditions during the storms which

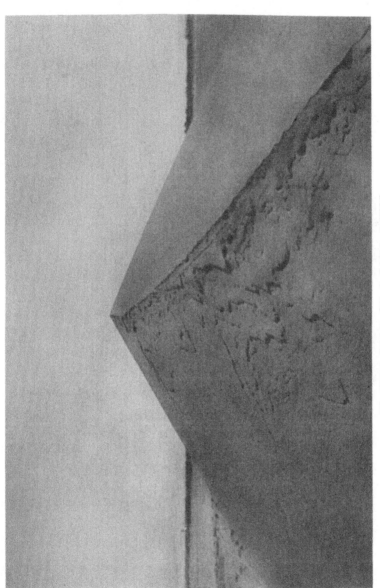

12. THE PEAK OF A SAND DUNE
(Ismailia, Egypt)

blow over sand-seas. When the air is electrical the power of the wind to throw the sand into suspension is no doubt enormously increased. I have not seen the process at work under natural conditions but I made some simple experiments and found that a slight charge imparted to sand contained in a basin made the sand-grains dance so rapidly that the air above the vessel presented the appearance of a haze, although there was no dust present with the sand.

(9) *Waves in drifting snow*

In 1899 I found by personal enquiry that Arctic explorers were uncertain whether travelling waves similar to those in loose sand were produced by the action of wind upon snow. Dr Nansen, however, whom I met at the International Geographical Congress at Berlin in the autumn of that year, told me that he thought they did exist although he could not be sure.

In the winter of 1899–1900 I went to Grantown-on-Spey in Scotland for a fortnight during a spell of severe weather in order to make observations on snow-drift, but, although the snow drifted freely upon the open moors, the flying particles were adhesive, and there was no formation of normal travelling waves with steeper slope to leeward.

I decided therefore to spend the winter of 1900–1 in Canada in order to investigate the question, partly on account of the facilities for travel provided by a transcontinental railway, partly because of the advantage of sunshine for observation and photography, which places within the Arctic circle do not enjoy during the season of really low temperature.

During December 1900 the snow in the neighbourhood of Montreal did not exhibit new or specially interesting forms, but one night in January the temperature suddenly dropped to *minus* 8 degrees Fahrenheit, forty degrees of frost, and next morning the appearance of the neighbouring countryside had greatly changed with the transformation of the snow from a damp, adhesive to a dry, slippery material. The sun shone brightly in a blue sky, but a strong wind was blowing, and the fresh snow, which had fallen to an average depth of 3 inches during the night, was flying freely, hazing the air to a height of some 30 feet. Upon the encrusted surface of the hardened snows of December lay a train of transverse waves of the new snow, travelling slowly but visibly, during the lulls as well as during the gusts of the wind. The measured advance in 40 minutes was 6·5 feet, a much greater speed than that of sand-waves.

These snow-waves had the normal desk-shaped profile, in some places with a rounded crest over which the snow poured in a cataract, in others with a knife-edge crest. From this the drift seemed to leap, but I could see a backward drift along the surface of the old hardened snow, towards the foot of the cliff.

An important difference between the snow-waves and the sand-waves of the Nile foreland near Helwan was the complete absence of rippling on the snow, owing perhaps to the extremely friable character of the material.

13. TRAVELLING SNOW WAVES
(Near Montreal)

The average distance between the crests of the ridges was 15 feet 10·4 inches, and the average height 4·9 inches, so that the ratio of length to height was 38·9, only half the steepness of sub-aerial sand-waves.

In the afternoon of the same day I came to a place where a group of travelling snow-waves was forming, their wave-length being 5½ feet, and during the course of the winter I never came across any of smaller size.

The important fact that no upstanding obstacle and no inequality of underlying surface is necessary to originate a group of such waves was well illustrated by their occurrence upon great stretches of smooth, un-broken ice on the frozen St Lawrence. Here the waves were unusually perfect, for there was no hard surface exposed between the ridges but, on the contrary, deep, soft snow below the troughs.

My next place of sojourn was Winnipeg, where the surrounding prairie is level and open. On January 23rd, 1901, in a moderate gale of wind, the snow-drift on the prairie was a remarkable sight, only the roofs and chimneys of the farmhouses being visible above the flying particles. Nevertheless no waves were formed, for there were only eight degrees of frost and the snow was adhesive, the flakes plastering every snowy surface which faced the wind.

On the 18th of February, however, after a fresh fall of snow 5 inches deep, there were 28 degrees of frost, the snow was dry, the particles did not adhere to one another, and waves of snow diversified the whole ex-panse of the prairie, all travelling in a slow, ghostly procession before a strong breeze. Each wave sloped gently upwards to a nearly level summit which terminated in a sharp edge, beyond which was a short, steep slope of 30 degrees. The crest-length was many times greater than the wave-length. An unbroken series of twenty-nine ridges had an average wave-length of 30 feet and height of 7·2 inches, the ratio of height to length being, therefore, 50.

When the march of the waves had continued for some hours, I measured an uninterrupted series of one hundred and ten, and found that the wave-length had increased to 32 feet 10 inches. The most notable change, however, was an increasing irregularity of wave-length, for whereas the average difference between one wave-length and the next had been at first 22 per cent. of the mean, it was now 36·5 per cent. One cause of this deterioration of the pattern was evidently the consolidation of the snow by pressure, whereby when the lower layers of the ridges became exposed their material was no longer of the perfectly granular kind. At night, when the wind usually dropped, the surface of these waves became hard and compact, owing, as I surmised, to evaporation from the earth-warmed lower layers and condensation in the interstices of the surfaces, where the cold of space was encountered. Thus both from compacting below by pressure and by consolidation by cold at the surface, the travelling snow-waves were prevented from growing to a size comparable to that of desert dunes.

A fresh breeze was sufficient to keep the loose, dry snow in continuous suspension, hazing the air to a height

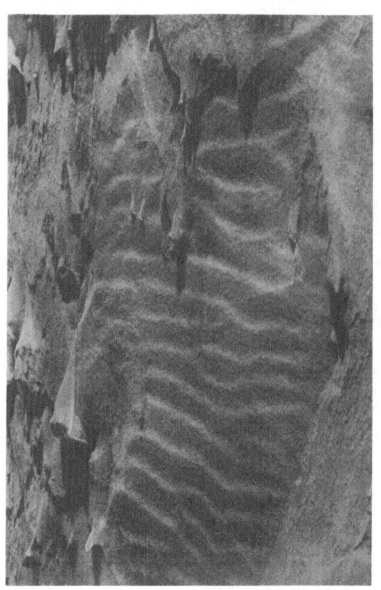

14. RIPPLES IN GRANULAR SNOW
(Winnipeg)

of about 30 feet and forming travelling waves of a size not to be confused with ripples. A breeze of similar strength dealt in the same way with the very fine and filamentous river sand of the Nile, but only on two occasions have I seen the beginning of the process in ordinary sand (coarser, and more rounded), once in the desert and once on the shore.

The experiments of Dr J. S. Owens, of which I shall presently give further particulars, have proved that when water is highly charged with sand the rate of subsidence of the particles is diminished. Thus when a certain critical velocity is reached the water suddenly becomes turbid, for deposition diminishes at the same time that erosion increases. This is also a critical velocity for wave-making, since every incipient hollow will be deepened and every incipient hillock heightened.

One of the results of the diminished rate of deposition in turbid water is the formation of a definite surface of separation between the turbid and the clear. It is significant that a similar condition occurs in the drifting of dry sand by wind, for, as Professor W. H. Hobbs has pointed out, the sand-blast of the desert erodes masonry and polishes iron posts up to a perfectly well-defined level of about 3 feet above ground.[1] This fact points to the conclusion that the rate of settlement of granular material in air is less when the air is thick with flying particles than when there are comparatively few particles in suspension. If this conclusion be correct, as I think

[1] See "The Origin of Desert Depressions", *Annals of the Association of American Geographers*, Vol. VII, 1918, pp. 25-6.

it is, the formation of aeolian waves (as distinguished from ripples of small height comparable to the diameter of the granules) will begin suddenly at a critical velocity which will differ for different materials according to their rate of settlement.

As loose, light, dry snow has less power than sand to resist the lifting action of the wind it stands in waves of less steepness. Similarly, when the snow is drifting

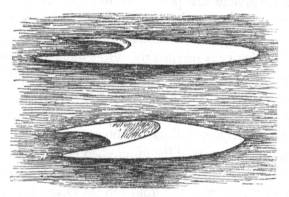

Crescentic dunes in snow (above) and sand (below).

in isolated patches, the crescentic mounds into which it is piled are not so steep as the Barchans or Medaños of the desert. These snow crescents, moreover, are narrower in proportion to their length, i.e. have finer lines in the horizontal as well as in the vertical plane.

Although the drifting waves of fresh-fallen snow were unrippled, there was a kind of snow-grit which fell into ripples resembling the sub-aerial sand-ripples. I kept a look-out for the possible formation of this material by consolidation due to pressure, but saw no evidence of the process and I infer therefore that it is produced by the breaking-up of a frozen crust by the action of wind.

In these ripples, as in sand, the grains were coarser on the crests. A group in which the average size of the grains was about $\frac{1}{32}$nd of an inch had an average wave-length of 2 inches. This was twenty-nine times the height, so that they were much flatter than sand-ripples, but steeper than waves of fresh-fallen snow. In one case I recorded an advance of 10 inches in 2 minutes during a fresh breeze, a much greater rate of advance than that of sand-ripples.

There were also patches of rippled snow-grit in which the grains were mostly between $\frac{1}{12}$th and $\frac{1}{8}$th of an inch in diameter with some as large as $\frac{1}{4}$th of an inch in the surface layer. These ripples were steeper, shorter-crested, and with an arched cross-section.

One day when I was on the prairie near Winnipeg the ripples in granular snow provided a remarkably instructive exhibition of the operation of the active eddy of the wind in the lee of a travelling surface wave. A shallow depression in the hardened snow, a few feet in diameter, and roughly circular, was being filled with fine, gritty particles which the breeze was driving along near the surface. The whole of this newly filled depression was soon covered by a uniform pattern of transverse ripples. Presently the gritty deposit piled up a little higher than the surrounding surface of hardened snow, and at once the weather side began to bank up above the rest of the rippled patch. No sooner was this excrescence formed, than the snow-grit along the central (down-

71

wind) line of the patch began to travel backwards along the surface. The ripples, hitherto sharply defined, became indistinct, and were soon obliterated over the whole width of the patch except at the extreme edges which had least shelter, for the ridge to windward was arched in cross-section. I have little doubt that here was made visible the very process by which the shorter waves of the sea become less and less conspicuous during the growth of the dominant wave, which produces a vortex or eddy whose longer sweep hinders the development of shorter waves.

The forms carved in the surface of snow compacted by pressure provided a striking illustration of the readiness with which wind falls into a regularly undulating flow. The initial stage of erosion was the formation of a group of transverse ridges with the steep side *facing* the wind, the length of the ridges being many times greater than the distance between them, although the front of each ridge was markedly sinuous. Here was a surface with the form of a group of waves. Moreover, as the wind eroded each cliff and whisked away the loose, non-adhesive particles, the whole group travelled after the manner of a group of waves, except that, having regard to the direction in which the ridges faced, the movement would be better described as a retreat than advance of the wave-form. When, however, we carry the critical consideration of the surface-form and its movement a little further, we find a fundamental difference between the progression of these wave-like surfaces and that of waves in loose, granular material, for

15. UNDULATING SURFACE OF SNOW PRODUCED BY WIND EROSION

(Winnipeg)

in the former there is no re-deposit of the drifting granules.

What we have here, therefore, are neither waves in the full, dynamic sense, nor a mere wave-form, as in an "undulating" landscape, but the mould or impress of a group of true waves, the super-incumbent waves of the air.

This eroded surface did not long preserve its transverse cliffs, for the wind, concentrating in the notches, eroded them more rapidly than the salients, so that after a time the transverse ridges were cut through and longitudinal ridges of symmetrical cross-section resulted, and these were the final forms of erosion. In the moist snow of the Highlands of Scotland I observed, however, a very singular exception to this rule. This was during the visit to Grantown-on-Spey, to which I have already referred. There was a moderate breeze blowing, which was eroding the soft deposit of snow, forming a regular pattern, wave-like in appearance, of little transverse cliffs a fraction of an inch high, facing the wind. These receded slowly, but visibly, during the gusts. Snow was falling all the time, and at each sudden lull the whole pattern leapt up-wind, and, although the length of the jump was less than an inch, its suddenness made the occurrence spectacular.

There is, I think, no doubt as to what happened. During the gusts, the falling flakes were eddying about on the up-wind side of the little cliffs which, as I have said, faced the wind. Then, at the lull (which always came quite suddenly) two things occurred, first, the

73

wind ceased to carve away the little cliffs and, secondly, the falling flakes, which were adhesive, plastered the faces of the cliffs, building them out, so that all the cliffs jumped up-wind simultaneously.

(10) *Sub-aqueous sand-waves*

The Dorset coast between Bournemouth and Poole Harbour where I first observed the ocean swell and ripples in wind-driven sand provided also my first opportunity of studying sub-aqueous sand-waves. I began to study these in 1896 and found them so interesting that I followed up the subject by visits to localities on the coasts and estuaries of England, Wales and Scotland where large stretches of sand are covered and uncovered by the tide.

I found that a little stream flowing through a sandy beach at the slow rate of from half-a-foot per second left the water clear but made inequalities in the underlying sand of which the most noticeable feature was a little cliff, using that word to denote a steep ramp not an escarpment. When the depth of water in the shallower parts was less than 1 inch these little cliffs were not exactly transverse to the general direction of the stream nor separated by regular intervals. In streams of the same slowness but greater depth, however, the cliffs were transverse to the current and arranged with a sufficient approximation to uniformity of distance to look like a group of waves, and their movement down-stream, keeping station with each other, confirmed the indication given by their appearance. In such a stream on the

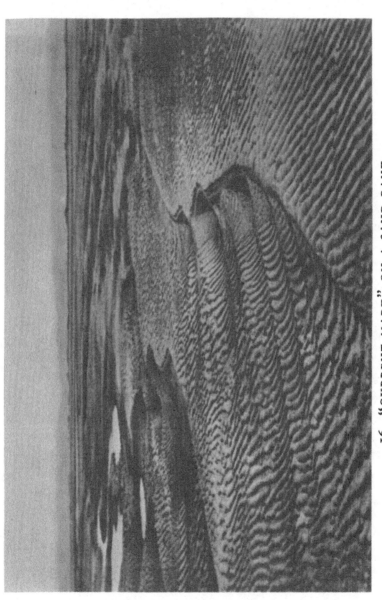

16. "CURRENT MARK" ON A SAND BANK

(Aberdovey)

sandy beach at Mundesley in Norfolk, flowing at half-a-foot per second, and with a depth of 5 to 6 inches. the wave-length of the sub-aqueous ridges was $5\frac{1}{2}$ inches. I observed that from a certain position to leeward of the deepest places the grains of sand departed in two directions, some travelling up-stream towards the foot of the preceding cliff, some down-stream towards the brow of the next. Thus the smallness of scale and the gentleness of the motion made it possible to see in operation a process which is, no doubt, fundamental in the formation, growth and movement of sub-aerial and sub-aqueous waves in granular material, and which I have long regarded as important in the development of ocean waves by wind.

In this small stream no sand could be seen in suspension except a little which was churned up in the pool on the lee of each small cliff.

Although the depth of water over the nearly flat summits of the ridges was sufficient to save them from the deformation caused by local acceleration of current, yet the winding course which the stream pursued, as does every stream flowing naturally through loose or friable material, was unfavourable to perfect regularity of the ground plan, or pattern. This follows from the fact, long since shown experimentally by the late Professor James Thomson,[1] that at every bend the direction of the current at the bottom differs from that at the surface, whereas wave-making conditions are only perfect when

[1] "On the Winding of Rivers in Alluvial Plains", *Proc. Roy. Soc.* 1876 and 1877.

upper and lower current are in the same vertical plane. I observed, however, a remarkably regular series of little transverse sand-waves in the sandy bed of a stream which had been made to run straight and between vertical banks in order that it should conform to the ornamental requirements of a public pleasure garden. The straight channel was 7 feet wide, the depth of water $3\frac{3}{4}$ inches, the surface perfectly smooth, the rate of flow 1 foot per second. The bottom was entirely covered with sand, which presented an unbroken sequence of little cliffs facing down-stream, at regular intervals, accurately transverse and extending almost to the bank on either side. An unbroken series of forty-six ripples had an average wave-length of 6 inches and height of about 0·35 inch. I marked the position of the cliffs by driving in small sticks, intending to record the rate of advance, but at the end of a quarter of an hour there was no change of position. In a natural, winding stream of the same rate of flow the progression of the sand-cliffs was 7 inches in a quarter of an hour. Examining minutely the bottom of the canalised stream, I could see no disturbance of the sand-grains. The nearly level crests were remarkably smooth, and no individual grain projected above the closely packed yellow surface. The bottom of the pool on the lee side of each cliff was covered loosely by some coarse-grained foreign material of dark brown colour which also lay unmoved, in spite of the gentle pumping action of the water in the pool. It appeared that in this straight channel a condition of equilibrium had been reached, that the bottom had been modelled

76

to a permanent form, and that the form was either stationary or moved at a very slow rate.

I return now to the observations made in the little stream which, in the distant date to which I refer, made its way across the sandy beach at Branksome Chine. When the speed was as much as 1·5 feet per second the very shallow water was usually thrown into surface waves which carved the bottom into undulations of the same length and not very different profile. The group of water-waves worked its way slowly down-stream, owing, as it seemed, to the erosion of the banks at the places of origin, and the submerged sand-waves moved with them. The water being quite clear, I could watch the sand-grains travelling over the ridges, which were depleted on the up-stream, and built out in the down-stream, side.

Where, however, the slope of the bottom was greater and the speed of the stream reached about $2\frac{1}{4}$ feet per second everything changed in a moment. The water became turbid throughout with flying sand and assumed an entirely different wave-surface. In place of ridges extending diagonally from the sides and intersecting at the middle, the surface of the water was in symmetrical, rounded undulations extending all the way across the stream, at right angles to the current. Under each water-wave was a symmetrical sand-ridge of rounded form. The surface of separation between the watery sand and sandy water was sufficiently definite to permit of measurement.

The following is a typical example of the dimensions

of the waves. The wave-length of the sand-waves and superposed water-waves was 9 inches. Leaving out of account the slight slope of the stream, the rise of the sand-wave from trough to crest was 0·75 inch. The rise of water from trough to crest was 1 inch. The depth of water above sand at trough was 1 inch, at crest 1·25 inches, the amplitude of the water-wave being therefore 1 inch, of the sand-wave 0·75 inch, and the crest of the sand-wave 0·25 inch lower than the trough of the water-wave.

The surface of a shallow stream flowing at this speed over pebbles is thrown into waves, but their wave-fronts have an arrowhead pattern similar to the wave-track left by water-fowl when swimming, or by a ship on her course; and the transversality of the water-waves in the stream turbid with flying sand can, I think, only be accounted for by the supposition that a waved surface formed in the sand impresses a waved form upon the surface of the shallow stream. These streams often flow flush with the surface of the sandy beach or shoal, so that there is no obstruction of banks to originate the sand-waves, and the bottom is commonly free from obstructions.

Dr J. S. Owens, in the course of a beautiful experimental research, undertaken, I am happy to think, at my suggestion,[1] showed that when varying quantities of fine sand are shaken up with water in a tall glass vessel, the rate of subsidence of the individual sand-grain

[1] See "Experiments on the Settlement of Solids in Water", by Dr John S. Owens, *Geographical Journal*, Vol. xxxvii, Jan. 1911, p. 73.

depends upon the quantity of sand in suspension. When the water is almost clear the sand-grain settles quickly, when the water is turbid the sand-grain settles slowly.

It follows from this observation that when the water of a stream with a sandy bed is highly charged with sand-grains in suspension, the slightest increase of the rate of erosion will be accompanied by a diminution in the rate of silting, and *vice versa*, so that every little hollow will be deepened and every little hillock raised, and so a wave-surface will quickly replace the smooth surface.

In the little streams with which we are now concerned these waves attained a very remarkable size considering the small depth of the water.

Their most singular character, however, was that they neither remained stationary, nor drifted down-stream, but steadily progressed *up-stream* at a visible rate of motion. I watched an individual sand-ridge moving up-stream at the rate of nearly half an inch per second until it reached the head of the group. Here the water was flowing with a smooth surface and the sand-ridge was at once planed off flat. The next ridge then arrived from down-stream and was quickly planed off; and so the procession continued endlessly. This planing down of the sand-ridges may suggest a doubt whether they were not created by the water-waves, but I think that the process is better explained by the supposition that up-stream of the point of origin of the sand-waves the rate of flow of the water was below the critical velocity, and so levelled out inequalities in the ordinary manner of a current whether of water or air, which is merely rolling loose material along the surface, not throwing it into suspension.

I shall presently show that in deep streams with a smooth surface and a velocity of 2·5 feet per second or more flowing over sand-banks, the bottom is thrown into sand-waves, the height of which may be measured in feet, which move *down-stream*. It is evident, therefore, that the up-stream motion of the sand-waves in the little stream had to do with the shallowness of the water.

Observation showed that there was no eddy on the lee of the ridges, and it seemed that the sand came down everywhere in a slanting shower, as hail falls, so that the up-stream slope of each ridge received a heavier shower than the down-stream slope. This difference is sufficient to account for an up-stream movement. The absence of a vortex on the lee side of the ridges is, I infer, due to the stirring of the water from top to bottom by an immense number of sand-grains cutting across the lines of flow.

In June 1900 I spent three weeks at Aberdovey in North Wales in order to study the larger kind of sub-aqueous sand-waves. The estuary of the river Dovey revealed at low tide broad stretches of clean sand through which the land-waters made their way in a winding course similar to the meanders of a river flowing across an alluvial plain. The low-water exit was a narrow channel between sand-banks, but when the returning tide had risen above the level of these its waters flowed up the estuary in a broad sheet. The hills which rise steeply from the right bank or shore of the estuary provided a bird's-eye view of the expanse of sand-banks at low water, a marvellous pattern of sand-waves. These had the familiar desk-shape. The direction of the steep,

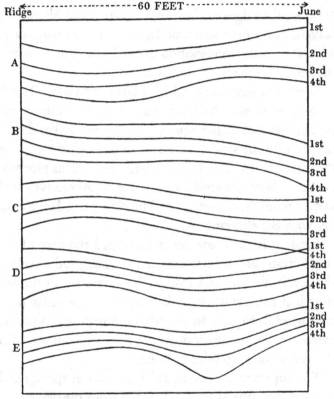

Ridge ---------------------60 FEET-------------------→ June
1st
2nd
A 3rd
4th

B 1st
2nd
3rd
4th
1st
C 2nd
3rd
1st
4th
2nd
D 3rd
4th
1st
2nd
3rd
4th
E

The movement of a group of tidal sand-waves (Aberdovey).

lee face indicated the direction of the current by which they had been formed. Those which faced inland, and must therefore have been formed by the current of the flood tide, had long, straight fronts when far from the low-water channels, but near these channels were more sinuous. In the former position, as many as fifty consecutive ridges without break of pattern could be counted.

The boatman of whom I made enquiries did not know whether the ridges on the sand-bank opposite Aberdovey travelled under the action of the tides or were stationary; all that ordinary observation proved was that their size varied with the strength of the tides.

Upon this sand-bank, which bears the name of Traeth Malgwyn on the Ordnance map, I marked out an experimental plot 60 feet square with twenty stakes driven deeply into the sand in five rows marking five cliffs of the sand-waves, four stakes in each row. This plot was on the side of the sand-bank nearer to Aberdovey and not far from the ebb-channel. The ridges faced seawards, i.e. with the ebb-current.

Landing here at low water I mapped the plan of the ridge-crests daily by reference to the stakes, and recorded the level of crests and troughs by reference to threads stretched from stake to stake. The measurements were continued from June 1st to 17th, both inclusive, a period extending from one set of spring tides to the next, with the period of neap tides between.

The approximate depths at high water at springs and neaps at this time of year, calculated from the tide tables, were respectively $9\frac{1}{2}$ and $6\frac{1}{2}$ feet.

The measurements on June 2nd, 3rd and 4th, while the spring tides were diminishing, showed a practically constant wave-length of about 13·5 feet, but the height diminished from 6·3 to 4 inches. No definite change

17. TIDAL SAND WAVES
(Aberdovey)

could be found in the mean surface level. Meanwhile the five ridges advanced as a group, keeping station with each other, the rate of advance being from $2\frac{1}{2}$ to 3 feet per day. Thus the total advance in three days was about two-thirds of the wave-length. Then on June 5th the movement of the ridges almost ceased, the advance being only $1\frac{1}{2}$ inches. Thereafter, until June 11th, the ridges lost not only height but sharpness of outline and regularity of front. On June 12th, however, following an increase of tide, the "experimental plot" was once more in sharply defined ridge and furrow. By June 15th the height of the ridges was 9·7 inches, and wave-length 11 feet 9 inches. The ratio of length to height was therefore 14·5, whereas on June 2nd, towards the end of spring tides, the length was 13 feet 10·4 inches and height 6·34 inches, the ratio of length to height being therefore 26·2. The comparison is instructive as indicating that the length is dependent on height, as would be indeed expected, since the strength of the vortex to leeward of the cliff is dependent upon the height.

On June 16th, the day of strongest tide, I determined the speed of the ebb-current when the depth of water over the sand-bank was 3 feet, and found that it was 3 feet per second. This is well above the velocity at which sand goes into eddying suspension. The depth above the sand-level was sufficient to maintain a smooth water-surface, so that the conditions in which these sand-waves were formed and driven down-stream resembled more nearly the conditions under which sub-aerial waves of sand or snow are produced than those of the sub-aqueous sand-ripples travelling up-stream in water so shallow that its surface is forced into a switch-back form.

Opposite Barmouth, on the sand-banks near the entrance of the Mawddach estuary, a group of eleven sand-waves ranged from 21 to 34 feet in wave-length. The greatest individual height was 2 feet 2 inches. The greatest depth of water at spring tides, according to the tide tables, is about 9 feet.

In the tidal estuary of the South Esk near Montrose, a series of ten sand-ridges ranged from 13 to 27 feet in wave-length with an average of 19 feet. Their steepness was of the same order as in the Mawddach and Dovey estuaries. In the course of two tides, the first ridge of the group advanced 30 inches and the last ridge of the group 31 inches.

All three estuaries are of the bottle-neck shape which ensures that the tide shall cover and uncover the shoals gently, and flow in a swift stream when the water is deep. These are evidently the conditions favourable both to the formation of large sand-waves and to their preservation on the dried-out sand-banks.

The greatest wave-length of tidal sand-ridges which I have measured was on the North Goodwins, where three consecutive ridges had an average wave-length of 72 feet. They were, however, much flattened. They faced with the flood tide, and the reason for this could be inferred from the fact that the part of the shoal to the north dried out before the ebb-current set in, so that these ridges were protected from the action of the ebb. The question arises, what would be the greatest possible height of these ridges before they were partly smoothed

84

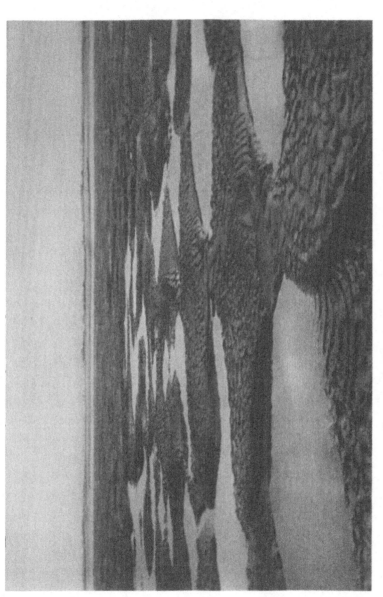

18. INTERSECTING TIDAL SAND WAVES
(Dun Sands, River Severn)

owing to the falling off of the tides or some other cause? Taking 14 as the maximum ratio of length to height for the sand-waves found on dried-out shoals, the height of the sand-ridges on the North Goodwins before they were impaired would have been 5 feet.

The sand-banks of the steeply sloping and funnel-shaped estuary of the Severn between Awre Point, below Newnham, and Beachley Point, at the junction of the Wye, are subject to a strong current during the last of the ebb, and so are mostly left with a nearly smooth surface when dry. Below Beachley Point, however, the rocky shoal called English Stones, projecting from the east shore, restricts the channel to a narrow passage during the latter part of the ebb and the early part of the flood. Below Beachley Point and above English Stones, within the eddy of the flood caused by the latter, is the shoal called Dun Sands. Landing here at low water, I found a sand-surface as strongly waved as the shoals of bottle-necked estuaries. The surface, however, presented a character which I had not met with in the estuaries of the Dovey, Mawddach and South Esk, for there were two intersecting sets of ridges of about equal size. One set faced down the estuary, the other diagonally across towards the low-water channel to which I have referred. The boatman who rowed me over informed me that the direction of the ebbing current changed suddenly when the "Stones" dried out, as is indeed inevitable. The formation of a pattern of intersecting ridges is of interest as illustrating the true wave-character of these sand-forms.

CHAPTER III

TIDAL BORES AND OTHER PROGRESSIVE WAVES IN RIVERS

(11) *On the discharge of a progressive from a standing wave in the Rapids of Niagara*

The waves of rivers differ in their mode of origin from the storm waves of the sea and those in sand and snow, for they are not inter-surface waves formed and driven by a super-incumbent fluid; and it is on account of this difference that I have postponed their consideration until after that of waves in sand and snow.

The ordinary undulations of tideless rivers are those called "Standing" waves, because they maintain their station relatively to the observer on the bank. The cusped form of a leading member of the group, and the boldly rounded outline of the succeeding undulations, are attractive features of the swift torrents of the mountain, and of those, as Niagara River, which cut their way from the plateau to the lowland.

In the neighbourhood of some projection from the bottom or the bank by which the rate of flow is retarded, the water rises in a crest from which the momentum of the descending current carries it below the ordinary level, forming the trough of the wave, and so on in undulations of equal wave-length but diminishing amplitude. The pattern, or surface plan, has a regular but elaborate symmetry. The complete, bi-lateral form is seen where some rock on the bottom makes a shoal in mid-stream. Here the water rises in a bowed front, and on either side the succession of hollows and crests spreads laterally, the echelon pattern having an evident likeness to that of the diverging wave-track formed by water-fowl when swimming, or by a steamer plying on a lake.

The train of standing waves which arises from re-sistance at the bank has, of necessity, only a uni-lateral arrangement, the group resembling the echelon waves on either port or starboard of a steamer. When the stream is narrow the waves which diverge from the opposite banks meet and cross in the middle, where the resulting, stationary mounds of water are sometimes higher than either component, even at the place of origin by the bank.

The current of a river never has an absolutely steady speed. In torrents the fluctuation is particularly marked, and to this I attribute a fluctuation of the standing waves, most marked in the case of the cusped wave which is usually at or near the head of the group. At each lull of the current the cusped wave moves momentarily up-stream and breaks at the crest. The occurrence is repeated at frequent intervals, imparting variety to the appearance of the wave and also a cadence to the sound of the stream. The fluctuation of the waves in most rapids is, however, very slight. My general impression is that the excursion is less than the breadth of the mound of water, so that the waves although not quite steady seem to be tethered.

Most rapids have a fairly straight channel, it is the rivers of alluvial plains which flow windingly. It happens, however, that one of the swiftest of the world's larger rapids has a very sharp bend in its course. This is the Middle Rapid of Niagara River which comes between the Falls and the Whirlpool. Here a remarkable kind of travelling wave is associated with the ordinary stationary waves.

The rapid begins where the Railway Bridges cross the Niagara gorge. Its length from the Bridges to the entrance of the Whirlpool is a little more than 1 mile. At the beginning of the rapid the channel narrows to about 300 feet and is reduced in depth from 160 to not more than 50 feet.[1] The gradient of the bottom suddenly steepens, and the river descends 50 feet in a distance of little more than 1 mile. In this reach a speed of 30 miles an hour has been recorded.

A light railway, built for the purpose of viewing the rapid, follows the right bank in the narrow space between the river and the cliff. This is the concave bank of the channel. A Station or Halt with a small observation platform is provided at the elbow of the torrent in order that the visitor may have the opportunity of observing at leisure the most striking feature of the most spectacular of the three rapids of Niagara River. This is the great Leaping Wave. The standing waves which extend diagonally across-and-down stream from either bank are superposed near the middle, and, as the rapid

[1] G. K. Gilbert, *Explanation of United States Geological Survey Map of Niagara River and Vicinity.*

88

19. LEAPING WAVE
(The Middle Rapid, Niagara)

is narrow, the elongated mound of water where the two crests over-ride is higher than the individual waves even at their places of origin.

Descending to the rocky margin of the torrent, I watched the great mound from the nearest and lowest attainable standpoint. It waxed and waned in height continuously and visibly, varying, as I estimated, from 15 to 20 feet above the ordinary level, and at the same time swaying to and fro in the current. Then, at short but irregular intervals, its waters leaped high into the air and flew down the gorge with terrific speed in a deluge of spray. The standing wave which remained was much diminished in height, but it was only a matter of seconds before it regained its ordinary dimensions.

The spectacle presented by the heaving, swaying and leaping of the wave, the coursing flakes of foam, and flying spray, was so fascinating in its tumultuous force and speed that I forgot all else, and was nearly caught by a surge of water which suddenly rushed in upon the rocks where I stood. Withdrawing to a position a few feet higher above the torrent, I soon saw that this inrush of water was merely an exceptionally large member of a roughly periodic surge which came in diagonally from the up-stream side. The intervals between successive surges were 20, 12, 15, 15 and 20 seconds.

The Middle Rapid is overlooked by the cliffs on either side which rise precipitously to a height of about 250 feet above the level of the torrent. From the edge of the cliff on the left, or Canadian side, I looked down on the waves at the elbow of the rapids. The standing ridges extending diagonally down-stream from the banks, and piled to double height where they met near the middle, were spread out as a map.

While I was looking at one of the ridges near the right bank, a strange thing happened. The ridge threw off a travelling wave, of which the form and orientation were the same as those of the parent wave, and this travelling ridge made its way across the torrent to the opposite shore, maintaining a crest on the up-stream side but drifting bodily with the current.

The height of this progressive wave was considerable, not indeed equal to that of the parent wave but of the same order of magnitude.

The standing waves near the left bank gave off similar progressive waves which made their way across stream, buffeting through those which came from the right bank.

These travelling waves were of particular interest in relation to the appearances which had so riveted my attention when looking from the water-level at the Rapids Railway Station. From my new position I saw the great standing wave near the centre of the stream, opposite to the Rapids Railway Station, struck simultaneously from the right hand and the left by travelling waves originating from the opposite sides of the torrent, whereon the great billow leapt up and flung masses of flying foam into the air. On one occasion half a dozen waves arrived so nearly together that all participated in the conflict and disruption. Such appears to be the mechanism of the Leaping Waves which present so

20. THE MIDDLE RAPID, NIAGARA, FROM ABOVE

fine a spectacle from the bank at the elbow of the rapid.

But although the Leaping Wave is the most spectacular appearance, the problem of geophysical interest is the origin of the travelling waves. This I take to be an alternate throttle and release of the headlong current at the elbow of the channel.

General experience of phenomena of the kind indicates that the river would become congested gradually and that the congestion would be relieved suddenly. The disengagement of a travelling wave from the standing wave, which is crested in a diagonal direction up-and-across stream, must, as I understand the matter, be attributed to the current suddenly becoming slower than the wave.

But the question naturally arises—how can the release of water *retard* the current?

I suggest that the explanation is to be sought in the well-known fact that the upper and lower layers at the bend of a river do not flow in the same direction, the upper layers impinging more directly upon the outer bank, the lower layers flowing more towards the inner bank. It is fairly evident that the release of the congestion of water at the bend of the Middle Rapid of Niagara River must be achieved by an increased flow of the lower layers which flow more directly towards the exit, and I suggest that this is attended by a diminished flow in the upper layers, so that the speed of the waves is excessive and they charge forward.

(12) *Roll waves in shallow water flowing in a flat-bottomed channel*

After the melting of the snows, the delightful valley of Grindelwald is filled with the sound of many waters; of the torrent from the glacier brawling among the boulders, of cascades racing in narrow channels through the steep pastures of the mountain side, and of water-falls, each of which has a cadence of its own.

If the rapid currents be viewed between narrowed eyelids, so as to blur the too insistent outline of stationary objects without unduly darkening the field, it is often possible to detect a travelling disturbance which out-strips the foam-flakes; something swifter than the stream and therefore in the nature of a progressive wave.

In torrents which are deep all the way across, or which vary from deep to shallow in cross-section, surface in-equalities travelling more swiftly than the stream neither grow individually nor acquire regularity of form or sequence. Where on the contrary the shallowest streams flow over slabs of rock, these inequalities acquire the form of cross-stream ridges. Above the lip of the lofty but slender water-fall of the Tsinglebach, near Bur-glauenen, in the gorge which leads from Grindelwald to Interlaken, there is a sloping slab of rock from which the little *roll waves* are launched into space in a succession of bars of foam, each of which is drawn out into an arrow-head in the descent, and so the water-fall is resolved into a processional cascade of white comets.

When staying at Montreux I had the opportunity of observing systematically the mode of flow of water on an artificial surface which might be likened to the slab of sloping rock above the Tsinglebach falls, but immensely prolonged. This was the cement floor of a little conduit of level cross-section beside the funicular railway which connects Territet with Glion. The conduit is 14 inches wide, has a length of 2250 feet and a maximum gradient of 1 in 1·75, and carries the overflow of water from the machinery of the funicular. Within a few yards from the summit at Glion, the flickering flow of the very shallow current is transformed into an orderly procession of small, progressive waves with straight, frothing fronts, ranged squarely across channel. I found the depth at crest to be 0·2 inch and at trough 0·1 inch. The space between the wave-crests steadily increased until a maximum of about 2 feet was attained. This was long before the end of the run, and thereafter the procession underwent no further change.

Once when the car in which I was travelling passed the other car of the funicular, a small quantity of water was discharged into the conduit from the passing car, and both the height of the roll waves and their distance apart were immediately increased.

On another occasion, however, when a large quantity of water was poured into the conduit on the arrival of our car at the summit station, the roll waves vanished, and were replaced by stationary waves extending from each side diagonally down-and-across stream. As the excess of water ran off, the shallowing stream at this place became unsteady, and then, after a few moments of flickering, resumed its usual state of periodic, gushing flow.

Many of the torrents of the Alps are carried in the lower part of their course through paved conduits, as a protection against flooding; and in these, when the water does not exceed a few inches in depth, roll waves are a normal occurrence. During the summer of 1904, which I spent by the shore of the lake of Thun, I made daily observations of the roll waves in the stone conduit which receives the Grünnbach torrent as it emerges from the gorge of the Justisthal and carries it for a distance of 450 yards to the lake, with a minimum slope of about 1 in 14. The conduit is about 15 feet wide, with nearly vertical walls and a flat cross-section. The floor is paved with squared and trimmed blocks, but their surfaces are not quite flush with one another at the junctions, so that the profile of the conduit floor has a succession of slight inequalities. These exercise an effect upon the flow which is just perceptible to the eye, but the fact which I have already recorded that roll waves are spontaneously developed on a smoothly cemented floor is a warning not to regard the palpable inequalities as prime factors in their formation.

Between the last pool of the natural torrent and the beginning of the flat-bottomed channel, the water flows for 50 feet over a slightly hollowed pavement, and here the current flickers. It then enters the 450-yard channel of the flat-bottomed conduit. Within the space of a few yards, something like a procession of waves appears,

21. ROLL WAVES, LOOKING UP STREAM
(The Grünnbach, Merligen)

little ridges passing the observer at the rate of about 120 to the minute. The depth here was usually about 2 inches throughout the almost uniformly dry weather of this summer.

At 150 yards from the entrance I observed notably larger waves, but there were also many small ones, and the general effect was still somewhat confused. Forty yards further on there was no longer any confusion, but an orderly procession of roll waves passing the observer at the rate of 33 to the minute; and for the remaining 260 yards of the conduit the height of the wave-crests and the distance between them continued to increase, and all were foaming.

At the end of the conduit the waves succeeded one another at the rate of 17 to the minute, but it is evident that the conduit was not long enough to allow the waves to attain a uniform maximum size. This is illustrated by the following observation of intervals between the arrival of the roll waves at the outfall, which varied from 1 to 10 seconds:

INTERVALS BETWEEN ARRIVAL OF ROLL WAVES AT
THE OUTFALL OF THE GRÜNNBACH CONDUIT
IN SECONDS

5,	3,	4,	8,	4,	7,	4,	4,	2,	5
4,	5,	6,	4,	5,	3,	3,	4,	1,	6
3,	4,	2,	4,	4,	3,	1,	7,	2,	7
9,	2,	9,	8,	1,	10,	1,	5,	6,	1
4,	9,	1,	6,	1,	2,	6,	7,	4,	5
4,	4,	4,	8						

The growth of the waves was achieved in great part by a process which could be easily seen. The higher ridges travelled more quickly than the lower, yet did not pass through after the manner of waves in deep water, but, on the contrary, incorporated the smaller members; so that, at each overtaking, a small ridge was obliterated and the large ridge increased in height, and therewith in speed also.

On August 20th I made a note at the time that I could see clearly that the depth decreased continuously from one crest to the foot of the next, and that the speed of the current also decreased continuously, so that *the stream was all wave.*

On September 16th, 1904, when there was more water than usual in the Grünnbach, namely about 3 to $3\frac{1}{2}$ inches at the entrance, the wave development was very striking. By the end of the run the depth at the crest was about 8 inches and at the trough $1\frac{1}{2}$ inches. On this day a slackening of the current just before the arrival of the crest was very noticeable.

The ordinary speed of the current was about 10 feet per second. The amount by which the speed of the roll waves exceeded the speed of the current was as follows:

ROLL WAVES IN THE GRÜNNBACH CONDUIT

Date	Depth at crest (inches)	Depth at trough (inches)	Wave-velocity (feet per second)
August 26th, 1904	2·5	1·0	2·06
September 8th, 1904	4·0	1·0	3·275
June 15th, 1904	4·5	2·0	3·54

I should add that the measurements are of the approximate character permissible in pioneer field-work, not of

96

22. ROLL WAVES, LOOKING DOWN STREAM
(The Grünnbach, Merligen)

the precise kind which could be obtained by the experimental installation of an hydraulic engineer.

It is of great interest to compare these results with the velocities recorded by the late Mr John Scott Russell, F.R.S.,[1] for a solitary wave mechanically generated in shallow, still water contained in a rectangular trough.

J. SCOTT RUSSELL'S EXPERIMENTAL DETERMINATION
OF THE VELOCITY OF A SOLITARY WAVE IN
STILL WATER

Depth at crest of wave (inches)	Wave-velocity (feet per second)
2·19	2·30
3·10	2·87
4·00	3·33
4·49	3·46

Scott Russell points out that these results are in accordance with the general formula for the velocity of tides and other long waves, expressed in feet and seconds, namely

$$V = \sqrt{gh},$$

where V is the velocity, g the acceleration of gravity, h the depth, *provided that the depth be reckoned from the crest of the wave.*

The general agreement of these results with the speed of roll waves in the rapid current of the Grünnbach is certainly striking when we consider the great difference of circumstances.

The genesis of roll waves in the conduit which carries the Guntenbach for the last 300 yards of its course to the

[1] *Report on Waves*, Report of meeting of the British Association at York in 1844 (published 1845).

lake of Thun differed considerably from that of the roll waves in the Grünnbach conduit. The depth of water was the same, generally about 2 inches, but the gradient, 1 in 22, was less steep, the sides were gently sloping instead of nearly vertical, and the stone pavement was of irregular, roughly hewn blocks. In the upper part of the conduit there were deep, longitudinal grooves in the paving, originating I think in the grinding action of the large stones carried down from the conglomerate cliffs of the Guntenschlucht. Here no roll waves formed. In the lower part of the conduit, where the pavement, although rough, was not scored longitudinally, roll waves formed suddenly at irregular intervals. Once formed, the roll wave grew rapidly in height and traversed the remainder of the distance to the outfall, never dying out by the way. The place of starting varied, and although I watched these occurrences almost daily for three months I never succeeded in seeing the start. In every case my attention was first attracted by a sudden sound (best imitated by that of the word "flop"). Turning my eyes quickly I saw the little roll wave already on its course.

On examination of the part of the conduit where the wave was first seen I usually found that it was near a slight depression of the pavement, extending across the whole width of the channel, forming a sort of pool. The most noticeable of these transverse depressions was at 120 yards from the outfall, and the wave started in this vicinity more often than at any other place.

The general depth of water was difficult to determine

23. ROLL WAVE, LEAPING OUTFALL
(The Grünnbach, Merligen)

on account of the roughness of the pavement, but was usually about 2 inches; rather deeper in the transverse pool and perhaps as little as $1\frac{1}{2}$ inches at the sill on the down-stream side. Thus a slight pulse in the current might be momentarily impounded in the pool and then burst in a wave over the sill. The process recalls the events in those Yorkshire streams the Ure, Swale and Tees, which make their way down from the moors through channels in the mountain limestone with remarkably regular cross-section, but presenting in longitudinal profile a succession of pools and sills. After thunderstorms or other heavy rain upon the moors, each of these rivers is liable to a roll wave of sufficient size and violence to be dangerous to anglers standing in the stream.

I found the speed of the current in the Guntenbach to be 8·5 feet per second and that of the passing wave to be 2·25 feet per second greater, which is concordant with Mr Scott Russell's measurements of a solitary wave in still water, within the limits imposed by the difficulty of measurement of depth in the roughly paved conduit.

The intervals between the waves at the outfall varied from 4 to 17 seconds, a much slower succession than in the steeper, better paved and nearly rectangular channel of the Grünnbach conduit.

On one occasion after heavy rain, when the Guntenbach conduit had a depth of 5 inches, no roll waves were formed.

In the spring of 1905, when at St Maurice in the upper Rhône valley, I came across a stone-paved conduit which was very similar in size and slope to that of the Gunten-bach, and with a similar depth of water, namely about 2 inches. The roll waves succeeded one another at intervals varying from 5 to 30 seconds.

The special interest of this conduit was due to the circumstance that the lower part was so encumbered with débris that the stream had to make its way through a winding channel with a cross-section varying greatly in depth, as in an ordinary, natural stream flowing through alluvium. The result was spectacular, for each roll wave in succession when it reached the place of unsymmetrical triangular cross-section instantaneously vanished.

The origin of roll waves in very shallow water in a steep channel of uniform cross-section, such as the conduit at Territet and that of the Grünnbach, can be inferred from the following considerations. The friction of the bottom greatly diminishes the rate of flow of a very shallow stream. Retardation of course increases the depth. But this increase of depth imparts a relatively great increase of velocity. Thus continuous flow is replaced by gushing flow, and the development is completed by the incorporation of the smaller and weaker by the larger and stronger gushes. In water of considerable depth the course of events is quite different, for the friction of the bottom has relatively little effect. Where local resistance retards the flow, forming a mound of water, the current runs slower, and a uniform discharge is maintained by swifter current at the troughs and slower at the crests of the switch-back surface of the resulting train of standing waves.

Although the explanation of the spontaneous evolution of roll waves in very shallow currents of uniform cross-section is simple, I am not aware that this phenomenon, which I began to investigate in 1904, had been predicted.

(13) *On tidal bores which assume the form of a group of short waves*

In most tidal rivers the first rise "cometh not with observation" but stealthily, while the current continues to run seawards, and the subsequent reversal of the current, although more noticeable, is not accompanied by any visible inflection of the water-surface.

There are, however, certain estuaries and rivers where the first rise of the tide with a greater spring which occurs near the new and full moon is a visible wave, sometimes solitary, sometimes the first member of a group. When solitary, the first rise and first flow are simultaneous, or very nearly so; when in the form of a group, the first flow follows immediately the rear of the group.

The local names given to this spectacle in different countries, and on different rivers in the same country, are in some cases certainly ancient, and their derivation is doubtful. In the eastern counties of England the occurrence is known as *eagre*; on the Garonne it is known as the *mascaret*. On the Seine the original name appears to have been *barre*, although the word *mascaret* is now alternative. The word *barre* is the more general term in French, and it is difficult to resist the conclusion that the English word "bore" has a common origin. The latter word is now invariably applied to the phenomenon on the Severn, and is used as an alternative for "eagre". This appears to be a change of fashion, for in Dr Johnson's Dictionary (3rd edition, 1765) the word "bore" does not occur in relation to tides, and the phenomenon on the Severn is referred to as the "eagre".

The word "bore" has now, however, become our general geographical term for the visible wave phenomena of the incoming tide in rivers in all parts of the world, as on the Petitcodiac, Amazon, Hugli and Tsien-Tang.

The word "bore" had also become a term of geophysics. Derived from the tidal phenomena in rivers, it is now also applied to the solitary, overfalling wave which succeeds the breaker upon flat, sandy shores. Whilst the word is thus extended to include a wave of different origin, its application in the case of tidal waves is commonly restricted by physicists to a particular variety of the natural phenomenon to which, as a word of common language, it was originally applied. The natural phenomenon diversely known as *eagre*, *mascaret* and *bore* or *barre* is not always, or even generally, a solitary wave; and in the Trent I know from personal observation that the solitary wave is not only exceptional but by no means the principal, or most spectacular, form of the bore. In this connection it is worth noting the divergence of the definitions of the "bore" in the

24. THE SEVERN BORE ON THE SHOAL
(Near Denny Pill)

New English Dictionary and of "*barre*" in E. Littré's *Dictionnaire de la langue Française*. The former has "a tidal wave of unusual height, caused by the rushing of the tide up a narrowing estuary"; the latter says "*La barre, les premières lames que la marée montante pousse dans un fleuve*".

The conception of the tidal bore which has become customary among English physicists must, I think, have been considerably influenced by the impression produced upon the late Sir George Airy, sometime Astronomer Royal, on the occasion when he viewed the Severn bore from the churchyard of Newnham in Gloucestershire, situated on a bluff which commands an extensive view of the river and estuary. I have myself seen the bore more than once from this vantage point, and it is indeed, as Sir George Airy wrote, "a majestic phenomenon".[1]

That the celebrated astronomer was profoundly moved by the sight is evident even in the restrained mode of expression characteristic of the scientific writing of his day. I think that some light will be thrown upon the subsequent treatment of the subject of tidal bores if we endeavour to enter into the mind of the observer on this historic occasion.

Sir George Airy had devoted himself to the problems of the moon's motion, and had investigated the tides of the world. His visit to Newnham was, apparently, rather for the purpose of witnessing the illustration of a phenomenon than in order to make comparative observa-

---

[1] *Encyc. Metropolitana*, article "Tides and Waves", by Sir George B. Airy. Afterwards published as a volume. Most English writers on the subject refer to this work, which evidently exercised a strong predisposing influence.

tions; and, as it seems to me, he was thrilled by the idea that what he saw was the advancing front of the great, solitary wave whose path across the oceans he had shown in his map of "cotidal lines". Such is now in fact the commonly accepted idea of a tidal bore.

When I first went down to the Severn to see the bore I was so fully imbued with this doctrine that I was slow to believe my eyes, and it was only after several repetitions of the phenomenon that I became convinced from observation that the visible inflection of the surface at the first rise of the incoming tide on the Severn is usually the front of a group of short waves, not of a solitary long wave. Later on I found that the eagre of the Trent is also normally a group of short waves.

An interesting commentary upon the discrepancy between the local facts and Airy's interpretation is afforded by certain remarks of the late Sir George Darwin,[1] the greatest authority of his day upon the theory of tides.

"In September 1897", he writes, "I was on the banks of the Severn at spring tide; but there was no proper bore, and only a succession of waves up-stream, and a rapid rise of water-level."

Thus it is evident that Sir George Darwin was on the look-out for a solitary, foaming ridge of water, and that no other kind of wave was in his view properly described as a bore. Yet the sentences which follow in his description show that he was not altogether easy in his mind as to the discrepancy between theory and observation.

[1] *The Tides, and kindred phenomena in the Solar System*, by George Howard Darwin (John Murray, 1898), p. 65.

25. THE SEVERN BORE IN THE POOL

(Near Denny Pill)

He writes: "... the heading back of the sea water by the natural current of the river and the progressive change of shape of a wave in shallow water combine to produce a rapid rise of the tide in rivers"; but adds that this "serves rather to explain a rapid rise than an absolutely sudden one".

The relation of height to length in tides is so small that a diagram has to be drawn with the vertical scale immensely exaggerated. I have not met with any record which indicates that either of these celebrated authorities upon the theory of tides examined the question as to the angle of slope which would be sufficient to enable the eye to detect an inflection of the water-surface.

The following observations, which I made on the river Trent in 1922 and 1928, explain the discrepancy between the observed, sudden, and the theoretic, gradual, rise which Sir George Darwin noted.

In 1922 I stayed from September 15th to 24th at Burton-upon-Stather, in Lincolnshire, a few miles from the outfall of the Trent into the Humber estuary, and made observations of the eagre up to a point a little beyond Wildsworth; and in 1928 I stayed at Gainsborough from August 16th to 19th and made observations from above Wildsworth to Torksey, which is nearly as far as the eagre runs.

On September 14th, 1922, the moon was at the last quarter, and the tides therefore in the neaps. New moon was on the morning of the 21st and the spring tides were of the great height which is common near the equinoxes.

On the afternoon of September 19th, two days before new moon, I was watching on the Jetty at Burton-upon-Stather nearly three miles above the outfall to the estuary. At 3.30 p.m. the current was running seawards at the rate of $2\frac{1}{2}$ feet per second, and I noted the level by means of a stone which was then awash. By 3.34 p.m. the stone was dry, and at 3.50 p.m. the level had fallen a very little lower. At 4.4 p.m. the stone was submerged, the level of the water having risen some two or three inches. The current, however, continued to run seawards, though very slowly, for another 4 minutes, that is to say, until 4.8 p.m.; but in the course of the next minute was reversed, the foam-spots at 4.9 p.m. travelling slowly up-stream. The interval between first rise and first flow was therefore 5 minutes, but at no moment was there any visible *inflection* of the water-surface.

On the 22nd, the day after new moon, I took my station on Garthorpe Jetty on the left bank facing the Burton-upon-Stather Jetty. At 6 p.m. the water was absolutely slack, so that the bore, which reached the Jetty 12 minutes later, was not headed back by any current. In very shallow water near the right bank there was a foaming ridge about 8 inches high which was a step in the water-levels, there being no depression behind; and the tide advanced over the mud flats under the left bank in the same manner. No sooner had the bore passed than the water was seen to be running up-stream.

At Keadby Bridge, $6\frac{3}{4}$ miles further up, where there are no mud flats, the appearance of the bore was very different, a long train of steep, rounded waves of which the first rose about 1 foot above the level of the river. The jointed masonry of the bridge piers provided a means of observing the rise of the water and of estimating its amount.

One hundred well-marked waves passed the pier in 3 minutes, by which time I judged that the mean level had risen 1 foot. The speed of the bore, measured from Flixborough, was 11 feet per second, so that as the period of the waves was 1·8 seconds the wave-length was 20 feet. The ordinary rule for calculating the maximum slope of deep-sea waves [1] is

$$180° \times \frac{\text{height}}{\text{length}},$$

which would make the maximum slope of the first wave 9°. Actually it would be considerably more, since the crest was advanced far beyond the middle point.

The general slope of the water in the group of one hundred waves was, however, only 1 in 2000, that is to say, an angle of less than 0° 2′. Even if the inequalities were smoothed out, the advance of such a slope would not, I presume, be a visible wave.

I made the following determination of rate of advance of the eagre and the period of its constituent waves:

Locality	Speed (m.p.h.)	Period (secs.)
Mere Dyke	9·0	2·3
Keadby Drain	8·0	2·0
Keadby Bridge	7·5	1·8
West Butterwick	9·9	2·0
Average	8·6	2·0

[1] See *A Manual of Naval Architecture*, by W. H. White (1896), p. 200.

The average length from crest to crest was therefore 25 feet.

At West Butterwick the height of the first ten waves was estimated at 4 feet, and at Wildsworth, where the river was narrower, the height of several waves was judged to be 5 feet, the highest which I observed at any place on the Trent.

Here the light was failing, so I did not pursue the eagre further but returned to my lodging at Burton-upon-Stather.

In 1928 I resumed work on the eagre of the Trent, choosing Gainsborough as my base, and began observations from a point a little above Wildsworth, following the eagre to Torksey which, as I have said, is near the end of its course.

On the afternoon of August 16th, 1928, the day after new moon, I estimated the height of the waves at Ravensfleet (not far from Wildsworth) at 3 feet. At Gainsborough Toll Bridge the height had increased to at least $3\frac{1}{2}$ feet.

On the morning of the 18th the height at Bowling Green Road, just below the town of Gainsborough, was estimated at $3\frac{1}{2}$ feet, but at Knaith it was only one-half of this height, and at Torksey it appeared to be rather less. At each of these places the bore was a group of many waves, the height of which from trough to crest diminished very gradually from front to rear; but where the bore was highest the graduation was more rapid, the most spectacular part of the disturbance being the first ten or fifteen waves. This character I had also noticed in 1922 at, and below, Wildsworth.

26. THE TRENT BORE
(Near Gainsborough, Lincs.)

At Torksey, on the morning of August 18th, 1928, an observation was made which confirmed and extended the lesson learned at Keadby Bridge. At Torksey a canal joins the Trent at right angles on the right bank. Leaving Knaith, where I had seen the bore arrive, I outstripped the tide in a motor car and stationed myself at Torksey at the angle between river and canal on the up-stream side of the canal, the river flowing by me, on the left, on a straight reach. Presently the bore appeared, half a mile away, at first as a broad, bright band advancing with a smooth and stately motion, then revealed as a train of rounded waves one foot and a half or less in height. When the bore passed the upstanding pro-montory between the river and canal (opposite to my post of observation) the procession of waves passed on up the river. I did not notice any such waves swing round the rectangular bend into the canal.

The entry of the tide into the canal was, however, very spectacular. The water gave one great heave, so slow that it seemed to have more in common with a long ocean swell than with the progressive waves of rivers. This slow rise I take to have been the true profile of the whole bore, freed from the usual corrugations. The great mound of water at the entrance of the very shallow canal ranged itself with a definite transverse front, and then three or four rounded swells extending from bank to bank, I think 20 feet at most from crest to crest, quickly formed, and travelled up the canal. After a run of about 350 yards these burst against the lock gates which bar the way. The resurgence also took the form of a group of waves athwart the canal, but although two or three front ridges were similar to those which had travelled up the canal the rearward part of the pattern was quite different, being composed of diagonal, not transverse ridges. The difference is no doubt due to the fact that the channel was being partially emptied, not filled as on a rising tide.

When the group of resurgent waves reached the exit of the canal the whole of the bore had passed on upstream, although still within sight until it rounded a bend of the river about 37 miles above the junction of the Trent with the Humber.

The last observation during 1928 was made at Ravensfleet on the morning of August 19th, four days after new moon, when the tides were decreasing. The front wave had decreased from the 3 feet of August 16th to $1\frac{1}{2}$ feet, as judged by the eye, and the general appearance of the group was very different, the processional effect being calm, instead of boisterous.

I counted forty waves in all, the first thirty being the most conspicuous. The time occupied in the passage of the forty waves was 1 minute 45 seconds. The wave-period was therefore 2·6 seconds.

On August 16th the most notable part of the disturbance here was composed of only eleven great waves, after which the members of the group were considerably smaller. The time which elapsed before smooth water was reached was 1 minute 40 seconds, which is nearly the same as on August 19th. The speed of the bore in the reach above Ravensfleet on August 19th was 13 feet per second, and, assuming this to have been the speed at Ravensfleet itself, the wave-length was 34 feet.

On this morning I saw a disturbance of the reflection some distance in front of the first wave-crest. On the day when the front wave was twice as high there was no indication of a rise of water at such a distance in front of the crest, and I judge therefore that when the tides fall off the front wave becomes more nearly symmetrical, the crest receding more and more towards the midway point. This is important as an indication of the rapidity of change from the spectacular to the invisible condition of the "first rise".

These observations might so far be taken to indicate that if the corrugations of the whole disturbance which constitutes the bore were smoothed out, the remaining long smooth slope would be the front of the true tide-wave, that is to say, the steepest part of a wave of which the rear would be many miles away. Certain peculiarities of the bore in the lower part of the tidal Trent suggest, however, that even this long slope may be but the front of one of those *seiches* in the tide, of which the existence is suggested by the "notches" in the curves automatically drawn by a tide-gauge.

On the 21st and 22nd of September 1922 there was a second bore following the first, a group of waves similar in appearance but not so high. The phenomenon is familiar to the dwellers by the river and is called "the second shove". I noticed that between Butterwick and Wildsworth the second bore diminished whilst the first increased.

The ferryman at Stockwith, higher up, told me that the second bore sometimes reached this place. A resident of Gainsborough informed me that it did not come as far as the latter town.

On September 22nd, 1922, the day after new moon, the time between the arrival of the first and second bore was

At Mere Dyke	3·5 minutes.
At Keadby Drain	5·0 ,,
At West Butterwick	3·5 ,,

On September 21st, 1922, at West Butterwick the interval between the arrival of the first and second bore was 3·0 minutes.

The waves of the second bore were decidedly lower than those of the first, but I think that they often exceeded 2 feet in height. The wave-length appeared to be the same, and where the second bore was well developed the number of waves was about the same as in the first bore.

Is the second surge an incident of the progress of the tide above the outfall of the Trent, or does it originate in the Humber? In the estuary just below the outfall is Whitton Sand, a drying shoal 2 miles long by a mile wide dividing the estuary at low water into the broad Whitton channel on the right, or Lincolnshire, shore, adjacent to the Trent, and the narrower channel on the left, or Yorkshire, shore by way of Faxfleet. The distance by the latter is not much greater, but shallower water would account for a difference of 3·5 minutes in the arrival of "the second shove".

Want of data obliges me to leave the matter here, but I suggest a further investigation of the phenomena of the bore of the Trent by that kind of team work which has now become fashionable in scientific, as in other spheres of activity. The high spring tides of the Autumnal Equinox, which are specially favourable to the investigation, occur within the Long Vacation of our Universities, and a fortnight in September would be sufficient for the work. The prime essential is the co-operation of a number of workers adequately equipped with instrumental means, who could make simultaneous observations at different points. Tide-gauges would be required both in the Trent and in the Humber estuary between Trent Falls and Whitton Ness.

At points selected for cinematograph pictures of the bore in the Trent, stakes with altitudes marked in bands should be set up, so that the oscillations of level during the passage of the short waves would be recorded on the film. Soundings for depth should be made before and after each tide at the points selected for observation. From the data so recorded the phenomena could be mathematically investigated with a completeness never yet attained.

It will not be irrelevant to add that the phenomena are so thrilling to watch, and that rural Lincolnshire has such charm in the summer season, that a party of young scientific men from one of the Universities would probably find the expedition very enjoyable.

The bore of the Severn from Stonebench, 4 miles below Gloucester, down to Denny Rock below Minsterworth, does not differ greatly in average speed and height from that of the Trent between Gainsborough and West Butterwick. The two rivers are of about the same breadth and both are free from sand-banks in these parts. The channel of the Severn has, however, one important feature which is not shared by that of the Trent, being crossed in several places by broad "benches" of sandstone rock which suddenly reduce the depth. At these permanent shallows the Severn bore presents the fine spectacle of a foaming breaker, and in consequence it is at such places as Denny Rock and Stonebench that sightseers congregate.

On September 29th, 1901, I was at Stonebench for the third tide after full moon. I watched the approach of the bore from the moment when it rounded the bend of the river 500 yards away, a group of many waves, keeping station as they came. I judged the front wave to be about 3 to $3\frac{1}{2}$ feet high in the middle of the river, and in places 4 or 5 feet high against the banks. The front was unbroken, but steep enough near the bank to throw a dark reflection forward upon the water. On reaching the "bench" of sandstone rock, however, where there is a sudden shoaling of the water, the front wave mounted, curled over in a perfect scroll and then collapsed in a rushing surge of foam, as a breaker on a shingle beach. This occurred where the depth was suddenly reduced by 3 feet, which at extreme low water would be a reduction of 50 per cent.

In the case of the Trent, whose course is not interrupted by rocky benches, one of the characteristics of the bore which impressed me most was the way in which the front wave held together, mile after mile, though often foaming at some point.

At Stonebench I had seen the change in the appearance of the bore as it passed from deeper to shallower water. Below Minsterworth, I stationed myself in a different position, standing opposite to the pool on the up-stream side of the broad shoal called Denny Rock. When the bore came round the bend of the river the steep wave in front, in places curling over, was almost the only visible disturbance from my low point of sight near the water-level, and even when near almost monopolised the attention of the eye. Being then new to these things I expected that the bore would preserve this appearance, and could scarcely believe my eyes when there was a sudden transformation into a group of gently rounded swells as the bore entered the deeper water in front of me.

These two types of bore are clearly shown in the photographs which I took at the time, which have been reproduced in many geographical text-books.

On a later occasion, October 30th, 1901, at Framilode, lower down the river, I noted other incidents accompanying the passage of the bore from shallower to deeper water which clearly indicated that, even taking the disturbance as a whole, it could not be regarded as the steep part of a wave many miles in length.

I watched the approach of the bore, which presented a crested front with a few ridges visible behind, and the rate of advance, which was quite steady, became clearly impressed upon the eye. Then, on a sudden, the front lost its steepness, and, simultaneously, new, rounded waves appeared at the rear, so that the group comprised many more individuals. More singular, however, was the circumstance that the rate of movement seemed suddenly to be retarded. The speed of the individual waves was, I presume, either the same as before or slightly greater, but the rate of advance of the procession as a whole was much reduced by the formation of new waves behind. The depth cannot, I think, have been sufficient to reduce the group-velocity to one half the individual velocity, which is the full effect in water of infinite depth. Another factor, however, presumably contributed to the sudden impression of slowing down, namely, the recession of the crests of the waves from a frontal position to nearly the middle point of the undulation.

By the Severn ferry at Newnham in Gloucestershire I once observed a curious effect which helps to explain the suddenness of change from the condition in which the first rise of tide eludes the eye to that in which it is very spectacular. The ferryman had warned me not to expect a bore, as, although the whole rise of tide was great, the tide had no "head". On the opposite shore I saw, indeed, a small wave running and breaking on the edge of the shelving sand, but in the deeper water on our side, the true right of the river, I did not see any wave coming. I was therefore startled when a stake which projected above the mud near by was covered instantaneously, I should say within the space of one second. There must therefore have been a wave several inches in height travelling up the river. Waves of as little as 3 or 4 inches high, such as the wind raises on ponds or small lakes, are quite conspicuous, because short wind-waves are steep. In the case of the group which makes up the tidal bore the wave-length is constant, being dependent only on the depth, and therefore the average steepness diminishes in the same proportion as the height. The length in the present case would be of the order of 30 or 40 feet. If, now, we consider the conditions of visibility of a wave in the case of the surface of muddy water, such as that of the tidal Trent or Severn, we can easily realise that in any position where the sky only is reflected, a slight departure of the surface from the horizontal could not be detected. Only where objects with small and precise detail of form, as trees and foliage, were reflected, would such slight disturbance be made visible.

On a previous occasion with a tide no higher, if as high, I had seen a larger bore at Newnham. On consulting the fishermen I found a consensus of opinion that, as there was so little "head", I should probably find that the tide was "squandering" itself in the Noose, that is to say, the expanse of sands beyond Hock Cliff where the channel changes from riverine to estuarine form. Accordingly I crossed the river early next morning and took station at the foot of Hock Cliff to watch the coming of the tide. The preceding summer had been a dry one, so that the flowing tides had the advantage over the ebbing tides, and the main channel was now tha which followed the chord of the arc near the right bank.

Under these conditions the bowed channel, which is the normal course of a strong ebbing current of land waters, is apt to be wholly or partly blocked at the lower end. The first appearance of the flood tide was, accordingly, in the shorter channel, a crested wave which I judged to be 2 or 2½ feet in height coming steadily on between steep banks. I was standing near the upper entrance to the now disused ebb channel, which curved away to the left. When the bore arrived opposite my station, a steep-fronted wave, similar in character, swung round into this channel and travelled down-stream, the height of the bore in the main channel, travelling on towards Newnham, being much diminished. Twenty-five minutes afterwards, the bowed channel by the left bank having now filled up, a small bore in the form of a low ridge of foaming water travelled from this channel across the broad sand-bank to the straight channel near the right bank, which was not yet quite full. The sand-banks were now completely covered. The broad expanse of water appeared to be devoid of current and remained slack for 4 minutes, when a steady flow set in towards Newnham, 26 minutes after the "head of the flood" had passed. Thus, as the fishermen surmised, the head of the tide had been squandered among the sand-banks.

It is interesting to note that the flood tide operates upon sand-banks so as to avoid the formation of a bore. It is after a wet season, when the ebb channel between Hock Cliff and Severn Bridge is deeply incised so that the tide comes up this way, that, according to local testimony, there is a large bore in the river between Newnham and Gloucester. Below Severn Bridge the flood tide has sufficient preponderance to keep an alternative channel open in all seasons; and here (again according to local testimony) no bore is formed, or at most one which has only a short run.

The account of the Hangchow bore in the river Tsien-Tang given by Commander Moore, R.N., is that of a solitary wave.[1] At Haining, where there is a sudden contraction of channel, the condition changing from estuarine to riverine, Commander Moore describes the bore as 8 to 11 feet high, extending in a nearly straight line across the river which is rather more than one statute mile in width, travelling between 12 and 13 knots, its front cascade of bubbling foam falling forward and pounding on itself. The slope of this travelling cascade was uniform at any particular part of the front, but varied in different places from 40° to 70°, being highest and steepest over the deep parts of the river. Currents were recorded on the flood tide of 7, 8, 9, 10, and in one case 11 knots. Commander Moore adds that "The Bore cannot be accurately described as a 'wave'. It is in no sense an undulation; nor is there any depression after it has passed. The same particles of water which rise with such a significant jump in the neighbourhood of Rambler Island, are precipitated over the vast bar of sand into the Tsien-Tang".

Mr Bell Dawson in his report upon the bore at

[1] *Report on the Bore of the Tsien-Tang Kiang*, by Commander Moore, R.N., H.M.S. *Rambler* (Admiralty Publication), 1888.

Moncton in the Petitcodiac River[1] records an average rate of 8·47 miles per hour, and adds that the current after the bore passes appears to have the same surface velocity. A gauge was set up, by means of which an accurate trace of the rise of water was obtained at each tide. This trace shows an initial wave of about 3 feet in height at the front edge of a long water-slope of about 2 feet per mile, punctuated here and there with a sudden rise of a few inches.

To sum up; it appears that the tide advances in a river as a solitary wave when the rise is very gradual or when it is very sudden and great. The first is the ordinary invisible tide, the second is the extreme form seen, for example, in the great Hangchow bore. Between these extremes there is a considerable range of moderate slopes of tidal front which cannot persist either as a plane surface or as a single curve when travelling in shallow water. This slope breaks up into a group of waves of which the length is not very different from that of waves of the same speed travelling in deep water; and to such a group heralding the incoming tide no less than to the solitary wave the name "eagre", *mascaret, barre,* or "bore" has been given locally from time immemorial. I submit that the observations which I have recorded show that in geophysics also the term "tidal bore" should not be restricted to the solitary wave but include the case where the head of the tide takes the form of a group of short waves.

[1] *Survey of Tides and Currents in Canadian Waters*, by W. Bell Dawson, C.E. (Canadian Government Publication), 1899.

ADDITIONAL NOTES

By HAROLD JEFFREYS

The following notes are intended to serve two purposes. The phenomena discussed by Dr Cornish are essentially hydrodynamical, and constitute a branch of physics. To understand them completely we need to examine them both from the observational and from the theoretical standpoint; our success is to be judged by the extent of agreement between the results of the two methods of approach. Dr Cornish has obtained a larger quantity of observational material than any other worker on the subject; and in the course of his remarks he makes a number of contributions to the theory, some quantitative and some qualitative. Nevertheless some further indication of the extent of the explanations available seems to be desirable. Some of his results may, I think, be fairly said to have been completely explained theoretically by modern hydrodynamics; others have been partly explained, though a residue remains to be considered by theoretical workers; while others are hardly explained at all. The explanations, so far as they exist, should I think be of interest to navigators, engineers, and others directly concerned with the natural phenomena; while so far as the phenomena remain unexplained it seems desirable to call the attention of theoretical investigators to them.

(1) *The relations between wave-velocity, wave-length, and period*

Classical hydrodynamics gives the following formula for the velocity of a wave on the surface of a liquid:

$$c^2 = \left(\frac{g}{\kappa} + \frac{T}{\rho}\kappa\right) \tanh \kappa h. \qquad \dots (1)$$

Here c is the velocity of the wave, and $2\pi/\kappa$ is the wave-length; g is the acceleration of gravity, T the surface tension, ρ the density, and h the depth. The function $\tanh \kappa h$ approaches unity if the depth exceeds about half the wave-length, so that in many phenomena we can simply replace it by unity; then the velocity is practically independent of the depth, and the formula becomes simply

$$c^2 = \frac{g}{\kappa} + \frac{T}{\rho}\kappa. \qquad \dots (2)$$

We further denote the period by $2\pi/\gamma$; then we have

$$\gamma = \kappa c. \qquad \dots (3)$$

With actual values for water we find the following values, in C.G.S. units:

κ	c (cm. per sec.)	γ	$2\pi/\kappa$ (cm.)	$2\pi/\gamma$ (sec.)
1	31·7	31·7	6·28	0·198
0·3	57·4	17·22	20·9	0·365
0·1	99·0	9·90	62·8	0·634
0·03	181	5·43	209	1·16
0·01	313	3·13	628	2·01
0·003	572	1·72	2090	3·65
0·001	990	0·99	6280	6·34
0·0003	1810	0·54	20900	11·6
0·0001	3130	0·31	62800	20·1

Thus a wave with a length from crest to crest of 209 metres should have a velocity of 18·1 metres per second and a period of 11·6 seconds. Within the range of this table the velocity and the period are nearly proportional to the square root of the wave-length. In British units a period of 10 seconds corresponds to a velocity of 51 feet per second and a wave-length of 510 feet.

For shorter waves κ is greater and the surface tension becomes more important in comparison with gravity. Actually the velocity is a minimum for a wave-length of about 1·8 cm.; Kelvin called longer waves than this "gravity waves" and shorter ones "ripples". This book is concerned entirely with gravity waves.

If the depth is a small fraction of the wave-length we can replace tanh κh approximately by κh; then

$$c^2 = \left(g + \frac{T\kappa^2}{\rho}\right) h, \qquad \dots (4)$$

and for waves more than a few centimetres in length c is nearly $\sqrt{(gh)}$. In general this is a good approximation when the water is more than about a centimetre deep and the wave-length is more than about four times the depth. Subject to these conditions the velocity depends on the depth alone and not on the wave-length. Such waves are known as "long waves"; waves such that tanh κh can be replaced by 1, on the other hand, may be called "deep-water waves" or "surface waves". For intermediate values of κh we must use the complete formula (1). On the other hand (3) is always true.

Physically the reason for the dependence of velocity on depth is as follows. In deep water, as a progressive wave passes, all the individual particles describe circular orbits, moving forward at the crests and backwards at the troughs. The radii of these circles decrease rapidly with depth; at a depth of half a wave-length the motion is only $\frac{1}{23}$ of what it is at the surface. This fact has been utilised by Vening Meinesz in his determinations of gravity at sea. These are made by means of a combination of pendulums, as on land; but the motion of the pendulum at sea is disturbed by that of the ship due to waves, and special precautions have to be taken to reduce this disturbance. One of these is to make the observations in a submarine, which is submerged to a considerable depth, where the amplitude of the wave motion is much less than at the surface.

The nature of the effect of depth is that at the bottom there is no vertical motion; water cannot flow through the sea-floor. This extra restriction makes no difference in deep water, since the motion is negligible anyhow at depths over half the wave-length; but in shallow water the effect of preventing vertical motion at a certain finite depth disturbs the whole character of the motion. The paths of the particles become ellipses with their shorter axes vertical; at the bottom they merely oscillate backwards and forwards horizontally. In long waves the vertical motion is small compared with the horizontal at all depths, even up to the surface. This is clearly seen in tidal currents, where the horizontal velocity may be several feet a second while the vertical one is a few feet in twelve hours.

(2) *Condition at a crest*

So long as the vertical displacement is a small fraction of the wave-length the displacements are simple harmonic functions of the time and the horizontal distance. In larger waves the theory requires some correction; it is found both by theory and by observation that the crests become sharper and the troughs flatter. It can be shown that with a large enough movement the crest becomes a sharp ridge with an angle of 120°.[1] This imposes an upper limit on the amplitude that can be associated with a given wave-length.

(3) *The generation of waves by wind*

The theory just outlined is derived from classical hydrodynamics, which rests on certain hypotheses about the nature of fluid motion that are not strictly true. But it can be shown that the inaccuracy introduced by these hypotheses is very small in a large number of hydrodynamical phenomena, and in particular in nearly all those of wave motion. The actual nature of the motion in nearly every case of wave motion is one of those permissible on the classical theory; but a theory that takes fuller account of the properties of real fluids makes further predictions possible. The classical theory treats the motion of the water as undisturbed by friction within the water itself and by forces acting on the water from

[1] Cf. Lamb, *Hydrodynamics*, 6th edition, 1932, p. 420.

outside (except in tidal theory, where it takes into account the attractions of the sun and moon on the water). The real fluid has internal friction (called viscosity) and is affected by stresses on its surface produced by the air. There is nothing in the classical theory to tell us *what* wave-lengths, if any, can be excited by a given wind-velocity; but the actual existence of some relation is obvious from the slightest experience.

The real fluid differs from the classical fluid in two respects. The classical fluid, as we have remarked, cannot flow through a solid boundary; but it can slip freely over it. A real fluid can neither penetrate nor slip over a solid boundary; the nearer we go to the solid the smaller becomes the velocity of the fluid with reference to the solid, tending to zero when we get sufficiently close. This is a very severe restriction. But in another respect the real fluid is less restricted than the classical one. In the classical fluid the force across any small plane surface within the fluid is simply a pressure at right angles to that surface. This is still true in a real fluid at rest. But in a real fluid in motion the force across such a plane element is not necessarily at right angles to it; there are also components in its plane depending on how the shape of the neighbouring fluid is changing. When we examine the consequences of these components we are led to the property of *viscosity*. The real fluid can change its shape, but the internal forces within it always tend to reduce the rate of change of shape; if the shape is changing there is a continual loss of mechanical energy, which is replaced by heat.

This dissipation of energy does not occur in the classical fluid.[1]

Now any wave involves change of shape and therefore loss of energy. Thus in a real fluid an advancing wave is dissipated and ultimately destroyed by viscosity. In ordinary cases of waves on water this effect is small in the sense that a wave can travel through many wavelengths without marked reduction of amplitude; but in a fluid of high viscosity, such as treacle, it is hardly possible to produce waves at all.

The problem of wave motion in a real fluid therefore takes a somewhat different aspect from what it presents in the classical theory. On the classical theory, if we have a train of simple harmonic waves, they can travel to any distance without loss of amplitude. In a real fluid they will lose amplitude unless there is some external force that provides them with new energy as fast as the existing energy is dissipated. Now in the generation of waves the amplitude does not decrease; it increases. A fuller theory of wave motion must therefore consider how this new energy arises, and in what circumstances it exceeds what is dissipated. It turns out that an explanation is available; but even an outline of it requires a little further discussion of the motion of real fluids.

It can be shown that however a fluid may be moving,

---

[1] If we try to combine the assumptions of no slip over the boundaries and purely normal pressure within the fluid, we find that it would be impossible to make two solids in contact with the same fluid move with different velocities; nor could any solid rotate.

the motion of a small element of it may be separated into three parts: first, a uniform motion without change of shape or orientation; second, changes of dimensions along each of three perpendicular directions; third, a rotation like that of a rigid body about a point. The three parts are called respectively *translation*, *strain*, and *vorticity*. The strain of course may be the same in all three directions; then it is a mere symmetrical expansion or contraction without change of shape. In a liquid, and often in a gas, however, this dilatation is negligible; it is important in sound waves and in bodies moving through fluids at velocities comparable with the velocity of sound, but not in any of the present problems. All the strains here considered are such as to involve no change of volume, and the fluids may be treated as incompressible. It is in the behaviour of vorticity that the real and classical fluids differ strikingly. In a classical fluid originally at rest we cannot produce vorticity anywhere by gravitational forces or the movement of solid boundaries with reference to the fluid; the motion everywhere is simply translation and change of shape without vorticity. Such a motion is called *irrotational*. In a real fluid, on the other hand, motion over solid boundaries produces vorticity instantly, and this vorticity proceeds to diffuse steadily into the interior of the fluid. In some cases, such as the flow of water through a fine tube, this effect dominates the whole character of the motion. But where the viscosity is small the vorticity takes a long time to spread through any considerable distance, and is practically confined to fluid that has passed close to the surface of a solid.[1] This is seen in most ordinary cases where a solid is pulled through water; there is strong vorticity in the wake, but hardly any elsewhere. Thus there are many phenomena in the motion of real fluids where the flow is practically irrotational except in very limited regions containing fluid that has passed near a solid. It is on this principle that the theory of the flight of aeroplanes and airships depends. Now there is a curious property of fluid flow over a curved surface, which we perhaps see most clearly in flow round a sphere.[2] The fluid on approaching the front of the sphere separates and spreads over the surface; but it does not join up again behind. It breaks away from the surface a little behind the centre and forms a broad turbulent wake. It is found that the existence of this wake is the most important factor in determining the resistance of the fluid to the motion of the sphere through it. Generally speaking converging flow in the lee of an obstacle does not take place to more than a very slight extent. Instead, the fluid that has spread to pass the obstacle at its widest part travels ahead, and the rear of the obstacle, except perhaps in the very earliest stages of the motion, is occupied by an eddy or a system of eddies.

This phenomenon is very noticeable in a river behind a projection from the bank, or on the top of a moving

[1] For a fuller account see *Proc. Roy. Soc.* A, Vol. cxxviii, 1930, pp. 376–93. A familiar example is a tea-leaf floating in a cup; we can turn the cup round again and again, but the leaf shows that the tea does not turn with it.
[2] See the illustrations given by Prandtl, *J. Roy. Aero. Soc.* 1927, especially Fig. 23.

the lee side. So in waves travelling with the wind the exposed side has a downward velocity and is exposed to the higher pressure from the air, while the sheltered side has an upward velocity and is exposed to the lower pressure. Consequently the tendency of the variation of pressure is to encourage the water to move downwards where it is already moving downwards, and upwards where it is already moving upwards. It therefore continually supplies the waves with new energy and tends to increase their amplitude. The principle is the same as in helping another person in a swing; we increase his amplitude of motion by pushing him when he is already moving away from us. If the supply of energy from the wind exceeds the dissipation by viscosity the waves will increase in size.

By making the appropriate changes in the above argument we can see easily that when the water as a whole is originally at rest the wind can supply energy only if the velocity of the waves is in the direction of the wind and less than the velocity of the wind. This imposes immediately a restriction on the possible wave-lengths, since too long a wave on deep water might have a greater velocity than the available wind. It is found in a more complete investigation[1] that the condition that the waves may grow is

$$c \, (V-c)^2 \geqslant \frac{4 \nu g \, (\rho - \rho')}{s \rho'}. \quad \ldots \ldots (5)$$

Here V is the velocity of the wind, c that of the waves, ν the kinematic viscosity of water, g gravity, ρ and ρ'

[1] *Proc. Roy. Soc.* A, Vol. cvii, 1925, p. 197; Vol. cx, 1926, p. 245.

motor vehicle. In the river the projection shoots the water into mid-stream, while behind it there is an *up-stream* current close to the bank. The front of a car or motor omnibus deflects the air upwards, so that the top is practically sheltered, the main current passing over-head; but sitting facing the front we often notice a draught on the back of our necks, showing that the air at that level is actually moving faster than the vehicle. When I cycle in hilly country against the wind I often find that in climbing a hill I have a helping wind behind me, the main wind being deflected overhead, so that I am actually riding in the lee eddy.

Now when a wind is blowing over a series of waves we have a state of affairs closely analogous to that just mentioned. On the leeward side of each crest there must be sheltering, or even an eddy producing a current in the reverse direction; but the main wind passes over the eddy and strikes with its full force on the windward side of the crest. The air is moving downwards before it strikes the water, but then is forced upwards; and this change of direction shows that the point where it turns upwards is one of specially high pressure. Hence we infer that when a wind blows over waves the pressure is higher on the slopes facing the wind than on those sheltered by the crests.

This is just what is required to explain how the energy of the wave-train is maintained. When the train moves a short distance in the direction of the wind, the trough advances a little and the level of the surface is lowered all along the exposed side. Similarly the surface rises on the densities of water and air respectively, and s a numerical coefficient, which I call the "sheltering co-efficient". The quantity on the right is essentially positive, so that the product on the left can exceed it only if c differs sufficiently both from o and from V. For a given wind-velocity, therefore, there can be only a limited range of possible wave-velocities. Indeed the value of c that makes the left side a maximum (for given V) is $\frac{1}{3}V$; so that unless

$$V^3 \geqslant \frac{27\nu g \,(\rho - \rho')}{s\rho'}, \qquad \ldots \ldots (6)$$

there will be no value of c at all that will satisfy (5). This relation determines the gentlest wind that can raise waves at all; for any stronger wind there will be a range of possible wave-velocities and therefore of wave-lengths.

All the quantities on the right are known except the number s. If then we find experimentally the smallest wind-velocity that leads to the formation of waves, and substitute it in (6), we can find s. It appears that this velocity is about 110 cm./sec., and corresponds to a value of s of about 0·27. The corresponding wave-velocity is $\frac{1}{3}$ of this, and implies a wave-length of 8·8 cm. This has not been measured accurately, but the actual waves raised by a wind just strong enough to produce waves at all certainly have a length in this neighbour-hood. It appears therefore that the sheltering theory accounts satisfactorily for the main features of wave formation in its early stages.

132

With stronger winds there is, as we have said, a range of possible wave-lengths; but there is also a range of possible directions of travel. Our fundamental formula for the wave-velocity, given the wave-length, supposes that the waves have long straight crests at right angles to the direction of travel, the height of the crest being uniform. I have called such waves "long-crested". But it is quite possible for the height of the crest to vary transversely. In a long-crested wave the elevation of the surface may be written

$$\zeta = a \cos (\kappa x - \gamma t), \qquad \ldots \ldots (7)$$

where x is the distance from some fixed line perpendicular to the direction of travel, t the time, and a the amplitude. But instead we may have

$$\zeta = a \cos (\kappa x - \gamma t) \cos \kappa' y, \qquad \ldots \ldots (8)$$

where y is the distance from another fixed line in the direction of travel and κ' is another constant. It is found that such a wave can travel unchanged; but the height of the crest is no longer uniform, vanishing at intervals π/κ'. It can actually be represented as the resultant of two waves of the form (7), but no longer travelling along the axis of x. Such a wave is called "short-crested". It is found that if the wind is just strong enough to raise waves at all, κ' must be zero for the waves produced; in other words the waves must be long-crested. This agrees with the facts. But stronger winds can and do raise short-crested waves. This explains the irregular appearance of the surface of the water in a strong wind.

133

It appears, however, that in these stronger winds there is a tendency for the longer and swifter waves to predominate, especially the swell, which is a long-crested wave. The full reason for this has not been obtained; but we may see that it is not surprising. Our energy considerations impose no limit to the height that a wave can grow to; but we have seen that the wave theory itself imposes such a limit. In its development by Michell[1] it shows that the crest becomes a wedge of angle 120° when the range of height (trough to crest) reaches about $\frac{1}{7}$ of the wave-length. Thus a wave 14 cm. in length could never attain a range of height over 2 cm. If the wind continues to pump energy into it, the crest merely curves over into the trough, producing "white horses". The longer the wave, however, the greater the height it can reach. If then the wind is able to produce waves of different lengths, the shortest reach their maximum height quickly, but the longer can continue to grow for a longer time and to a greater height. Thus when the waves have travelled a long distance, with the wind blowing over them all the time, the longest will tend to predominate, simply because they can store more energy.

The quantitative results given above refer wholly to waves on deep water. If the water is so shallow that the motion becomes appreciable at the bottom, the supply of energy remains the same for the same velocities and

[1] *Phil. Mag.* Vol. xxxvi, 1893, pp. 430–7. Useful summaries of this and many other investigations appear in H. Thorade, *Probleme der Wasserwellen*, 1931.

amplitude, but the dissipation becomes larger.[1] It seems that the wind-velocity needed to produce waves of given velocity on shallow water is greater than on deep water. On a puddle under a centimetre deep no waves are produced except by a strong wind, and when they are produced they are ripples only 1 or 2 cm. in length; no long waves are formed however strong the wind.

It will have been noticed that the foregoing argument depends at several points on the relative magnitude of various quantities, which can be assessed only after a mathematical investigation making use of laboratory determinations of these quantities. The equations of motion of a real fluid are accurately known; but it is seldom possible to solve them without approximation, and the valid approximations differ in different circumstances. The approximations made here, however, are valid in the ordinary cases of wave motion and its relation to wind.

(4) *The question of group-velocity*

In many problems of wave motion, where the velocity depends on the wave-length, a prominent part is played by the group-velocity

$$C = \frac{d\gamma}{d\kappa} = \frac{d\,(\kappa c)}{d\kappa} = c + \kappa\,\frac{dc}{d\kappa}. \qquad \ldots\ldots (9)$$

[1] In my 1926 paper mentioned above, p. 130, there is a mistake in equation (15); the rate of dissipation includes other terms than that quoted, one of them depending on the vorticity near the bottom. It appears that this term is not negligible when we try to allow for the effect of finite depth, and this allowance therefore needs re-examination.

In gravity waves on deep water, the group-velocity is half the wave-velocity. Its chief interest is in a motion produced at a point or along a line by a sudden disturbance. Then waves of a large range of periods are generated, and spread out at different speeds. If we fix our attention on any single wave (crest to crest), it moves forward with the wave-velocity corresponding to its length. But as it moves it continually changes its length and its period, and with them its velocity; an individual wave does not move with any constant velocity. We can, however, determine the wave-length associated with any assigned place and time; for if t is the time since the disturbance and x the distance from the place where the disturbance took place, then

$$x/t = C. \qquad \ldots \ldots (10)$$

C is in general different for different wave-lengths; we choose the wave-length that makes C equal to x/t; then this wave-length is that of the waves near distance x at time t. Thus a given period, wave-length, and wave-velocity appear to travel out at a constant rate, and this rate is the group-velocity. If we consider two different wave-lengths, they travel with different group-velocities, and therefore become further and further apart as the time increases. But the intermediate wave-lengths remain the same; hence there are a larger number of waves between two definite wave-lengths. The number of waves in the train therefore increases with the time. What actually happens in gravity waves, where the wave-velocity exceeds the group-velocity, is that the waves in front pass out of the group between two wave-lengths, and new waves form and enter the group from behind. The energy within the group remains the same.

If only two slightly different periods are present, we may consider a pair of wave-trains superposed, given by

$$\zeta = \tfrac{1}{2}a\{\cos(\gamma_1 t - \kappa_1 x) + \cos(\gamma_2 t - \kappa_2 x)\}$$
$$= a\cos(\gamma t - \kappa x)\cos(\delta t - \lambda x), \quad \ldots (11)$$

where

$$\gamma_1 = \gamma - \delta,\ \gamma_2 = \gamma + \delta;\ \kappa_1 = \kappa - \lambda,\ \kappa_2 = \kappa + \lambda. \quad \ldots (12)$$

We consider δ and λ as small, so that the second factor varies slowly in comparison with the first. Then this second factor vanishes when

$$\delta t = \tfrac{1}{2}(2n+1)\pi + \lambda x, \quad \ldots \ldots (13)$$

where n is an integer. At any distance, therefore, the amplitude of the motion varies in a period $2\pi/\delta$. If we consider the places where the amplitude vanishes, they appear to move forward with velocity δ/λ; for if we increase t by τ and x by $\delta\tau/\lambda$, we do not alter $\delta t - \kappa x$. But

$$\frac{\delta}{\lambda} = \frac{\gamma_2 - \gamma_1}{\kappa_2 - \kappa_1} = \frac{d\gamma}{d\kappa} = C, \quad \ldots \ldots (14)$$

when γ_1 and γ_2 are only slightly different; thus the places where the waves interfere and the observed amplitude is small advance with the group-velocity.

This principle arises in the experimental determination of the velocity of light. We are not dealing there with a simple wave-train, but with a series of flashes obtained by interrupting a train at regular intervals by the teeth of a rotating cog-wheel. We measure the time it takes a flash to travel round a circuit of known length. Now such a series of flashes can be represented by a superposition of wave-trains, giving a wave of varying amplitude such as we have just considered. If we measure the velocity of a place of zero amplitude, or one of maximum amplitude, we get the group-velocity.

Our question, then, is to what extent the group-velocity is important in water-waves raised by wind. It seems clear that in their formation we are concerned only with the wave-velocity. The motion of the air over the surface is determined by the velocity of the wind relative to the crests; and the crests move with the wave-velocity. But when the waves produced by a storm spread out into a calm region the energy must remain associated with the original periods, which spread out with the group-velocity. It follows that when we observe on our west coast a swell generated by a storm in the middle of the North Atlantic, the individual waves we see are not those actually produced in the storm. The latter have dashed on ahead of the main group and left their energy behind. If then we observe the period of the waves beating on the coast and calculate from it the wave-velocity, we shall be right; but the time taken by the swell to travel from the storm to the coast is twice what it would take if it travelled with the wave-velocity.

A short-crested wave may be considered as the resultant of two long-crested waves moving in directions with a finite angle between them. The whole of the waves produced by a wind can therefore be regarded as a combination of swells of different lengths, whose directions are grouped about that of the wind, but not all coinciding with it. But when they leave the region of the storm they travel in different directions. An observer at a distance large compared with the size of the storm therefore sees long-crested waves of different periods approaching him; all necessarily come from within the storm area and therefore from nearly the same direction. Thus at a distance the short-crested type of wave is less conspicuous than where it is actually generated, simply because it separates into two long-crested parts that do not go to the same place.

Outside a storm area the size of the swell seems to vary considerably from one wave to another. One roll of a ship may appear to be much larger than several before and after it. This can be understood if we recognise that the apparent swell is, for the above reason, the resultant of a number with different periods, which sometimes interfere and sometimes add up.

(5) *The equation of continuity and Bernoulli's equation*

The first of these equations says simply that if we take any definite region of space, the rate of decrease of the mass within it is equal to the rate of outflow from it. It is really the principle of conservation of mass. In a liquid and in the types of motion of air that are considered here the changes of density are negligible, and the principle amounts to saying that as much fluid flows out of an element as flows into it. The saying "Still waters run deep" is an elementary application of the equation. If we consider any two parallel cross-sections

139

of a river, one where it is broad and deep, the other where it is narrow and shallow, all the water that crosses the first must at some time or other also cross the second; and if the surface between them is not to rise or fall continually the rates of inflow and outflow must be equal. But these rates are simply the areas of the respective cross-sections multiplied by the mean velocities over the sections. Hence a large cross-section corresponds to a smaller velocity than a smaller one. If therefore a particular stretch of a river is exceptionally broad or deep the velocity there will be low; while any local narrowing or shallowing involves a local increase of velocity.

In a way this result appears paradoxical; we tend to think of an obstruction as causing the velocity to decrease, not to increase. But this is because we are using a different standard of comparison. If we fill up the river at its deeper parts so as to make its section uniform, we may or may not increase its velocity there; that depends on whether the surface remains at the same level and on whether the total flow remains the same. (The total flow in a river would hardly alter, but that in a pipe determined by a difference of pressure at the two ends would.) The distinction is perhaps seen most easily by considering a crowd trying to pass through a narrow doorway. At the back of the crowd the width is much greater than at the doorway, and whatever the velocity of the people may be at the moment of passing through, the velocity at the back is less than at the doorway. This is as much as we can infer from continuity alone. If we have a wider doorway and people of equal determination, the velocity of the individuals going through will still be much the same, but there will be more of them in a given time and the velocity of those at the back of the crowd will be greater. In fact *the equation of continuity expresses a comparison between velocities in different parts of the same system in some one régime of motion*; it says nothing by itself about what happens if we alter the system by changing the form of its boundaries.

If a stream has a local projection from the bank, extending only a small fraction of the width, the cross-section there is smaller than elsewhere and the average velocity greater. To determine the distribution of velocity over the section involves other considerations, and we can only state it here. The excess velocity is confined to the neighbourhood of the obstruction; in mid-stream the velocity is almost the same as above it.

Bernoulli's equation concerns the motion of any portion of a frictionless fluid. If we concentrate attention on a particular small piece of the fluid, it is acted upon by the pressure-difference between its front and rear. This in general gives a force in the direction of motion, which tends to change the velocity. The particle is also affected by the component of gravity in the direction of motion. For both reasons the velocity of the particle tends to change as it travels; but it is connected with the pressure and with gravity by the relation

$$\frac{p}{\rho} + \tfrac{1}{2}q^2 = gz + \text{a constant.} \quad \ldots\ldots (15)$$

Here p is the pressure, ρ the density, q the resultant velocity, g gravity, and z the depth below some fixed horizontal plane. We see that where the pressure is great or the depth small, the velocity is small: always provided we follow the same particle.

Since the velocity is great near any constriction, the pressure there is small. We shall have occasion to use this principle later. It actually gives the reason why a flag flaps and why a rising column of smoke from a chimney quickly becomes turbulent. Suppose we have a plane surface within a fluid, with fluid on both sides flowing past it. If it is undisturbed it may be able to remain plane. But if part of it is displaced slightly towards one side, we constrict the fluid on that side, increase the velocity and lower the pressure; on the other side the pressure is increased. Thus we now have a difference between the pressures on the two sides, tending to push the boundary and the matter near it in the direction of the displacement. Thus the changes of pressure resulting from any displacement of the boundary tend to increase the displacement, which therefore increases to a large amount.

It might seem that this effect should contribute to the formation of water-waves by wind. But the tendency of a displacement of a horizontal surface to increase is resisted by gravity and surface tension, and the velocity needed to generate waves is found on calculation to be about 6 metres a second, about six times what is observed, and the wave-length proves to be much smaller than the observed one. It seems therefore that while this tendency to instability exists, it is less important than the effect of sheltering.

It must be noticed that the truth of Bernoulli's equation depends on the absence of any other forces tending to alter the velocity of the particle, and in particular of friction. The latter is always present, but is unimportant in those regions of the fluid where the motion is irrotational.

(6) *The transport of sediments by streams*

It was stated above that in a fluid of small viscosity it will take a long time for vorticity to be diffused through any considerable distance from a solid boundary. We are familiar with the eddying motion that exists behind a solid of moderate size drawn through water. But actually this motion is not developed until the solid has been in motion for some time. If a sphere is suspended in water containing particles illuminated by a source of light, and suddenly set in motion, and if a photograph is taken immediately, it is found that the motions of the particles show no sign of eddies. Indeed the motion of the fluid that they exhibit is precisely the same as would exist if the classical conditions held right up to the solid. This verification of the classical solution in the real fluid is simply a result of the thinness of the layer containing vorticity for a short time after the motion starts.

Now so long as the classical conditions hold the pressure in the fluid, *ceteris paribus*, is greatest where the velocity is least, and conversely, by Bernoulli's equation. Consider, then, a solid body suddenly placed on the bottom of a stream. Where it touches the bottom the motion of the liquid is blocked, and the velocity is zero. But

143

above the solid the velocity is considerable, actually more than the mean velocity in the neighbourhood. Hence the pressure is greater below the solid than above it, and provides a force tending to lift the solid. If it is greater than the weight of the solid (after allowing for the buoyancy of the liquid), it will lift the solid off the bottom. But if the solid is denser than the liquid it will not stay off the bottom. The classical flow breaks down in time, vorticity is formed behind the solid, and a force arises accelerating the solid in the direction of the stream. If it reached the velocity of the stream in its neighbourhood the difference of velocity would no longer be available to produce an upward force, so that there would be nothing to hinder gravity from bringing the solid down again. Actually of course gravity overcomes the force due to the difference of velocity before this has quite vanished. So the solid descends and again strikes the bottom. The shock when this happens destroys most of the velocity of the solid down-stream, and the original conditions are restored. Thus the solid jumps off the bottom again. In a stream with a suitable velocity, therefore, a solid body that is not too large can be carried along the stream, its progress consisting of a series of jumps. This motion has been observed by G. K. Gilbert.[1]

According to observations by Sokolow, sand-grains with a diameter of 0·25 mm. are moved in this way by a wind of 4·5 to 6·7 metres per second. For grains of

[1] Cf. Jeffreys, *Proc. Camb. Phil. Soc.* Vol. xxv, 1929, pp. 272–6, and references given there.

other sizes the required velocity will be proportional to the square root of the diameter. In water, on account of the difference in density, the requisite velocity needs to be divided by about 40.

(7) *Waves in sand*

The theory of the formation and motion of waves in sand differs in principle from that of waves in water. In water the actual motion is an approximation to one that can exist in a frictionless liquid; a motion affecting a depth of the real fluid comparable with the wave-length can in fact travel without much change for several wave-lengths even without forces from the air to maintain it. In sand or snow, on the other hand, the motion consists entirely of the elevation, transport, and re-deposition of surface grains; the grains at depths more than a few diameters are strictly at rest. The conditions in the medium are entirely different from those in a water-wave; on this ground some writers have thought it illegitimate to use the word "waves" at all in relation to granular material. I see no practical advantage myself in such a restriction; it could serve only as a warning that the dynamics of a liquid cannot be applied to granular material, but I think few people would need such a warning. There is an obvious resemblance between a train of waves on water and a ripple-mark on sand; there is also a resemblance between the standing wave behind an obstruction in a river bed and the snow-drift formed in the lee of a hedge. If the word "waves" is not to be used in the latter cases we shall need another name; but I cannot see that any confusion can arise through using the same name in both cases. If we say "water-waves" and "waves in granular material" the dynamical distinction is adequately brought out.

The condition that a particle may be lifted off the surface can be inferred from general principles; it will be of the form

$$u^2 = \alpha \, \frac{\sigma - \rho}{\rho} \, gd, \qquad \ldots \ldots (16)$$

where u is the velocity of the current over the surface, σ the density of the grain, ρ that of the fluid, g gravity, and d the diameter of the grain. The numerical constant α depends on the shape of the grain and is best found by experiments such as Sokolow's.

The conditions during transport are highly complicated because they depend on the phenomenon of turbulence. The nature of this has been examined most thoroughly for motion in pipes. When water is forced slowly through a fine tube the motion at all points is parallel to the axis; the velocity is zero at the boundary and has a maximum in the centre. But at a certain critical velocity this regular motion breaks down. The velocity is no longer purely along the channel, but involves irregular motions across it, while at any place the velocity along the channel is no longer constant, but fluctuates irregularly about a mean. This condition is called *turbulence*. In a channel, which is what concerns us most here, turbulence arises when [1]

$$Q = 300\nu, \qquad \ldots \ldots (17)$$

[1] Jeffreys, *Phil. Mag.* Vol. XLIX, 1925, 793–807.

where Q is the quantity of fluid crossing unit width in unit time, and ν is the kinematic viscosity. For water ν is about 0·01 cm.²/sec., so that the transition comes when Q is about 3 cm.²/sec. In water with a mean velocity of 1 cm./sec. the depth cannot therefore exceed 3 cm. without the motion becoming turbulent. In all motions of rivers and even in quite shallow currents the motion of the water is therefore turbulent. In air ν is about 0·1 cm.²/sec., and again we find that turbulence should and does arise in all winds of ordinary velocities.

In viscous motion the tangential stress over the boundary is proportional to the local rate of shear; this remains true in turbulent motion, but an alternative empirical determination is available. If the mean velocity u is measured in the body of the fluid, the stress is about 0·002ρu^2. The numerical coefficient 0·002 does not depend on the units chosen so long as they are consistent; if for instance the stress is measured in dynes per square centimetre, and the velocity in centimetres per second, the density must be measured in grams per cubic centimetre. The viscosity does not appear in this equation; so long as the velocity is great enough to make (17) true, the viscosity hardly affects the frictional stress over the boundary, though it must be important in determining the details of the motion in the body of the fluid.

The reason for the transition from regular to turbulent motion has been the subject of mathematical investigations for about 50 years, from Osborne Reynolds onwards, but it cannot be said to be properly understood. We must at present regard it as simply an experimental fact. In perfectly smooth tubes the resistance at high velocities seems to be somewhat less than $0.002\rho u^2$, another factor arising depending on a low power of v/Q, perhaps $\frac{1}{4}$; but this does not affect rough boundaries such as those of granular material. Within the fluid the transport of momentum towards the boundary is much greater than in regular motion, because it is carried not only by viscosity, but also by the transverse movements involved in the eddies. It appears that a fair representation of the distribution of the mean velocity can be attained by introducing the idea of an "eddy-viscosity", which is larger than the ordinary viscosity, but enters the dynamical equations determining the mean motion in nearly the same way.[1] This eddy-viscosity, however, is not a characteristic property of the fluid; it depends on its state of motion, and it appears that even if we are dealing with different phenomena going on in the same fluid at the same time it has not necessarily the same value.[2]

We need not discuss turbulence in detail here, but we must notice that it plays an essential part in the transport of solid grains. When a grain is lifted off the bottom it is in general surrounded by rising air or water. So long as it is in an eddy current with an upward component, the resistance it offers to the current will tend to keep it up, and though it will rise more slowly than the fluid it may be thrown up to a considerable height before it

[1] G. I. Taylor, *Phil. Trans.* A, Vol. ccxv, 1915, pp. 1-26; *Proc. Roy. Soc.* A, Vol. cxxxv, 1932, pp. 678-702.
[2] Cf. L. F. Richardson, *Proc. Roy. Soc.* A, Vol. cx, 1926, pp. 709-37; Jeffreys, *Nature*, Dec. 26, 1931.

comes down again. The motion of the fluid itself is as much upwards as downwards, so that on an average a grain let loose in the interior of the fluid will descend under gravity. But this average statement cannot be applied to every individual grain, and a grain may have time to travel through several upward and downward currents before it again strikes the bottom. We cannot at present develop even a rough quantitative theory; but we shall expect that for a given size of grain the amount of solid in suspension and the average height that a grain reaches before it begins to descend will both increase with the strength of the wind.

Exner[1] gives the following empirical formula for the density of sand in air:

$$\gamma = (Ae^{mu} - B)e^{-rz}; \qquad \ldots \ldots (18)$$

here γ is the density of sand in grams per cubic metre, u the wind-velocity, z the height, and A, B, m, r constants to be found by experiment. He finds, for sand just capable of being lifted by a wind of 7 metres per second,

$$r = 11/\text{metre}, \quad m = 0.18 \text{ sec./metre},$$
$$A = 6.8 \text{ gm./metre}^3, \quad B = 22.5 \text{ gm./metre}^3.$$

He applies this rule to the migration of a sand-dune with uniform slopes on both sides. From the formula the total mass of sand being carried over the crest in unit time can be calculated; then supposing that it is deposited uniformly over the leeward side he finds how deeply it will cover this side, and hence how fast the dune will

[1] *Ergebnisse der kosmischen Physik*, Vol. I, 1931, pp. 373–445.

advance. A dune he examined should according to the theory travel at a rate of 15 metres a year; the observed rate was 7 metres a year. The agreement in order of magnitude indicates that the formula at any rate takes the chief factors into account.

He notices also that the rate of advance is inversely proportional to the height of the dune. This is because the rate of transport of sand over the top is independent of the height, whereas the area it has to cover on the lee side is proportional to the height. Thus a low dune will travel more rapidly than a high one. If a dune falls off in height towards its ends, the ends will advance faster than the middle, and the barchan form will be developed. The advance of the ends, however, causes the wind to cross them at an acute angle and to deflect the air along them without passing over the top. Thus deposition at the ends is reduced and the advance is slowed; the barchan thus approaches a steady form such that the whole of it advances at the same rate.

Exner's theory is an important contribution, but clearly does not explain the whole of the facts.[1] We want to know why a layer of sand, originally flat, develops inequalities of height at all, and what limits their height when they are formed and prevents them from growing indefinitely. The latter point may be understood if we refer again to the equation of continuity. The formation of a sand-hill means a constriction of the path of the air,

[1] I think the r in his formula (18) must decrease as u increases; a strong wind must throw grains to a greater average height than a weaker one.

so that the velocity must increase as the air ascends the windward slope. It is therefore greatest at the top and will raise the largest particles there. Thus erosion will be most severe at the top. It has often been noticed, originally by Sir G. H. Darwin, that a wave raises clouds of sand at the tops of all the ripples, while elsewhere the water remains perfectly clear. It is probably this extra velocity at the crests that limits the heights of dunes and sand-ripples.

The smoothness of the slope on both sides of a dune may be explained on the same lines as that of a scree slope at the foot of a cliff. A stone falling on such a slope may rebound several times before it comes to rest. Each stretch it descends implies a gain of energy from gravity, but each rebound implies a loss through imperfect elasticity and friction. If the former is on an average the greater, most of the material added at the top will shoot to the bottom and reduce the slope; if the latter is the greater, the stones will stay near the top and increase the slope. If the slope was vertical the gain from gravity would be the greater, since a stone could have at most one collision on the way down; if it was horizontal, the gain from gravity is absent and the stone comes to rest after a few turns. Thus there must be an intermediate slope such that the two effects balance. This is the slope taken up by the scree. When it has been attained any added matter will distribute itself uniformly.[1]

The sand thrown over the crest of a dune falls near the top on the lee side and slides down in much the same

[1] *Geol. Mag.* Vol. LXIX, 1932.

way as the stones on a scree; the only complication is the lee current blowing up the slope and thus resisting gravity. Thus a balance of energy for the average particle will occur only for a somewhat steeper slope than on a scree of the same material. On the windward side, however, the relations are reversed. It is essential to the transport of sand that the wind shall be strong enough to make the grains move along the flat ground; and with any extra speed it will be able to transport them uphill. In fact every particle being moved up the dune is being carried by the wind against gravity. The condition for the average particle when the form has become steady is therefore

> Energy gained in each flight due to the wind force − Energy lost through gravity = Loss of energy on rebound.

If the wind force is uniform, so will the slope be; but as we should expect the velocity to increase somewhat towards the top we should also expect the slope to increase somewhat. At the extreme top this relation will be reversed on account of the top being blown off.

The beginnings of any sand-wave seem to depend on turbulence. In an actual wind the velocity is not uniform, nor is the pressure at a given height. The effects of this can be seen even on water when the velocity is insufficient to generate waves. The surface in anything but an absolute calm has a slightly mottled or granular appearance, presumably due to these slight variations of pressure. A sudden gust gives a "cat's paw", a local region of intense mottling due to short-crested waves a few centimetres long; from this the regular train of waves emerges. Similar gustiness when the wind blows over sand will lead to local scouring and therefore to the formation of irregularities of the surface. Then any irregularity produces a lee eddy, so that air driving sand over it mixes with air from the lee eddy, loses some of its velocity, and drops the sand. Originally, therefore, we should expect small disturbances to increase with time. But it remains to be seen why they sometimes become long-crested wave-trains and sometimes isolated masses, and what it is that determines the wave-length that they actually adopt.

Many striking theoretical results about the formation and motion of sand-dunes, ripple-mark, sand-banks, and the meanders of streams have been given by Exner in the paper already mentioned. Meandering is shown to depend on very much the same processes as the formation of ripple-mark. When a stream silts up at one side, the sand-bank forms a lee eddy, with a vertical axis this time, and deflects the stream to the opposite bank with extra velocity, so that there is rapid erosion. This in time leaves the bank further down projecting; hence there is a lee eddy behind the new projection and further silting there. Thus silting and erosion occur alternately on each side and a river, originally straight, becomes sinuous.

It is sometimes said that meandering is a characteristic of slow streams. But I have never seen a lowland stream meander quite so much as the torrents that cross the peat on the slopes of Ingleborough. It seems that meandering cannot depend on slope or velocity alone, but must be influenced in some way by the nature of the bed.

(8) *Roll waves*

Some years ago I published[1] what appeared to be an explanation of the roll waves noticed by Dr Cornish in channels in Switzerland. I noticed similar waves myself in March 1928 in the outflow channel of a reservoir west of Oswestry. I am in some doubt, however, about the validity of the theory. It seemed to show that the surface of a stream in turbulent flow would cease to be regular at a slope of about 1 in 100; so far as observation goes roll waves do not develop at slopes under 1 in 10. Since the waves are really discontinuous in character they probably have some theoretical similarity to bores; but there is one fundamental difference, since the motion of any stream is controlled by friction, whereas bores probably occur in spite of friction.

(9) *Tidal bores*

The theory of the formation of bores consists of three parts, each satisfactory in representing one stage of the phenomenon, but the transitions between the stages when the three discussions hold remain uninvestigated.

The first stage is that of a tidal wave in a river or on a shelving shore, produced by the oscillation of the water-level in an open ocean or sea. In the ocean the

[1] *Phil. Mag.* 1925, *loc. cit.*

amplitude of the semidiurnal tide is probably only about a foot, but around coasts it may be 20 feet, and in some rivers even more. The chief cause of the increase in an estuary or a river is the narrowing of the channel as the wave proceeds upwards. The height of the tide at the mouth is determined by the moon's attraction and by the configuration of the ocean as a whole. The rise and fall of the water at the mouth necessarily produce periodic currents in the river. But this motion is of the nature of a long wave, that is, the wave-length is many times the depth of the water. If the water is 10 metres deep, the velocity of a long wave is 10 metres per second, or 430 km. in 12 hours. If the water is only 1 metre deep, the velocity is 140 km. in 12 hours. Hence in a long shallow river the wave-length of the tidal wave may easily be less than the length of the river, and the phase of the tide at different points of the river may differ by amounts up to 360°. If now the river was uniform in breadth and depth we should get considerable magnification of the tide if the natural period of the water in the river (say twice the time it takes a long wave to traverse its length) agreed with the period of the tide. But even without such agreement, which could in any case be a property of only a few rivers, we shall get magnification if the breadth and depth decrease as the distance from the mouth increases. The reason for the effect of breadth is that the energy of the inflowing water must be accommodated within the river. A sudden change of breadth gives a reflected wave; but with a gradual change the energy goes on almost without loss. The energy is proportional to the square of the amplitude and to the product of the breadth and the wave-length. It is found (originally by G. Green) that subject to the variations of breadth and depth being gradual (roughly, that they do not vary by large fractions of themselves within a wave-length) the amplitude of the tide varies as $b^{-\frac{1}{2}}h^{-\frac{1}{4}}$, where b is the breadth and h the mean depth across a transverse section.

As the tidal wave in the ocean approaches a shore, the variation of b does not arise, but that of h does; it is for this reason that tides on the coast are normally several times those in mid-ocean. In a river there will be further magnification.

Now in a shallow river a tide magnified in this way will have a height comparable with the undisturbed depth of the river, and one of the approximations of the wave theory fails. We can in fact show that when this happens the pure harmonic wave cannot travel unchanged. For a wave advancing up a river of uniform depth into undisturbed water it is found[1] that every elevation or depression of the surface travels with its own characteristic velocity $g^{\frac{1}{2}}(3\zeta^{\frac{1}{2}}-2h^{\frac{1}{2}})$, where h is the depth of the water not yet reached by the wave and ζ is the actual height of the surface above the bottom. For waves of small height ζ is nearly h, and the wave-velocity reduces to $(gh)^{\frac{1}{2}}$ as for long waves of small amplitude. But when the height is comparable with h, the crests of the waves move much faster than the troughs, and the waves therefore undergo great changes of form as they

[1] Lamb, *loc. cit.* art. 287.

156

advance. In fact when they have proceeded far enough the crests will actually overtake the troughs. Then there will be nothing to support the weight of the over-hanging water, and it must fall over into the trough. At this stage or earlier, therefore, the type of wave just considered must change into one such that the highest parts are chasing the lowest up the stream and continually overflowing them.

The next stage is naturally that of a bore. Here we have a quite sudden change of height at a certain part of the channel. An elevated mass of water moves up-stream, continually overflowing and absorbing the shallower water in front of it. This can be treated as a discontinuous wave. If the heights of the surface above the bottom on the high and low sides of the wave are h_1 and h_2, and the water on the shallow side is at rest, it is found[1] that the wave advances towards the shallow side with velocity $\{gh_2 (h_1+h_2)/2h_1\}^{\frac{1}{2}}$.

There seems to be no difference in principle between a tidal bore in a river and a breaker on the beach. Both depend on, first, the magnification of a wave in height owing to constriction of the boundaries; second, the fact that when the height has become comparable with the undisturbed depth of the water the crests proceed to overtake the troughs and fall over; and third, on the possibility of the forward propagation of a discontinuity of level.

The theory determining the velocity of a bore or a breaker, however, is obtained by simply considering the

[1] Lamb, *loc. cit.*

157

interchange of matter and momentum across two planes situated in front of and behind the wave itself. It does not consider the actual motion at the discontinuity, nor does it even prove that a discontinuity is essential to the phenomenon. If, in the second stage of the process, the water of the crests can begin to flow into the troughs before the wave has advanced so far that the crests are actually above the troughs, the bore may be merely a rapid rise of level and not an actual discontinuity. On a beach not all waves give breakers; and there is nothing in the theory inconsistent with the milder form of the bore that Dr Cornish describes in the Trent. The usual notion of the bore as an advancing wall of water is, however, exemplified in the remarkable photograph of one in the Tsien-Tang given by Thorade as Plate 4 of his book. There is a definite need of an extension of the theory to account for the difference between the forms of the bore observed in different cases.

Dr Cornish notes that when a sudden elevation enters deep water it is converted into a train of smooth waves. This would be expected. It is analogous to the train of waves generated when extra water is suddenly added at a point of the surface, or roughly to those produced when a stone is thrown in.

Friction is sometimes mentioned as contributory to the bore. I think that this is an error, and that friction actually tends to prevent it. Friction implies a continual loss of energy in any wave as it travels, and we should expect it, therefore, to oppose the tendency of the amplitude to increase in narrow channels and shallow water.

Actually in rivers affected by bores the bore dies out when it has proceeded a certain distance, in spite of the fact that the river is still becoming narrower and shallower. There seems to be no means of accounting for this decrease except by friction, and we should regard the bore as formed in spite of friction and not because of it.

INDEX

Printed in the United States
By Bookmasters